Electric Vehicles - Modelling and Simulations

Electric Vehicles - Modelling and Simulations

Editor

Helmut Krakowski

scitus
academics

Electric Vehicles - Modelling and Simulations
Edited by **Helmut Krakowski**

ISBN: 978-1-68117-414-3
Library of Congress Control Number: 2016937616

SCITUS Academics LLC,
www.scitusacademics.com
Box No. 4766, 616 Corporate Way,
Suite 2, Valley Cottage,
NY 10989

Notice

Printed in United States of America on Acid Free Paper

Preface

An electric vehicle (EV), also known as an electric motor vehicle, uses one or more motors for propulsion. Electric cars (EVs) are used only for electricity purposes. They are powered by one or more battery-powered electric motors. EVs have a number of advantages over internal motor vehicles for combustion. Electric vehicles convert about 59%–62% of electricity from grid to wheel power –conventional gasoline vehicles convert only about 17% – 21% of the energy stored in gasoline to wheel power. EVs do not emit tailpipe pollutants although electricity can be emitted from the power plant. The generation of electricity from nuclear, hydro, solar or wind energy does not cause air pollutants. Electric motors provide fast and quiet operation and require less maintenance than ICEs. However, researchers are working on improved battery technologies to improve driving distance and reduce charging time, weight and costs. In the end, these factors determine the future of electronic devices. EVs first appeared in the mid-19th century, when electricity was one of the preferred propulsion methods for motor vehicles, providing a level of comfort and ease of operation that gasoline vehicles could not achieve at the time. The internal combustion engine (ICE) has been the main propulsion method for motor vehicles for nearly 100 years, but electricity has remained common in other vehicle types, such as trains and smaller vehicles of all types.

This book focuses on the modeling and simulation of electric vehicles and their components. Mathematical models and components of electric vehicles have been introduced and combined to make this book a guide for industry, academia and policymakers.

Table of Contents

CHAPTER 1

A Comprehensive Analysis of Plug in Hybrid Electric Vehicles to Commercial Campus (V2C)

Andrew D. Clarke, Elham B. Makram

Department of Electrical and Computer Engineering,
Clemson University, Clemson, USA

ABSTRACT

Vehicle to grid is an emerging technology that utilizes plug in hybrid electric vehicle batteries to benefit electric utilities during times when the vehicle is parked and connected to the electric grid. In its current form however, vehicle to grid implementation poses many challenges that may not be easily overcome and many existing studies neglect critical aspects such as battery cost or driving profiles. The goal of this research is to ease some of these challenges by examining a vehicle to grid scenario on a university campus, as an example of a commercial campus, based on time of use electricity rates. An analysis of this scenario is conducted on a vehicle battery as well as a statio- nary battery for comparison. It is found that vehicle to campus and a stationary battery both have the potential to prove economical based on battery cost and electricity rates.

KEYWORDS

Electric Vehicles, Power Distribution, Economic Analysis, Vehicle to Building (V2B), Battery Storage

1. INTRODUCTION

Plug in hybrid electric vehicles (PHEVs) are becoming more prevalent. Many studies seek to examine possible services these vehicles may provide to electric utilities. In order to understand the implications of introducing PHEVs to a distribution system, one must first define what constitutes classification of a PHEV. Hybrid electric vehicles (HEVs) use both an internal combustion engine and an electric system usually consisting of a battery and a motor to meet the requirements of propulsion. A vehicle must meet the following three requirements beyond those of a typical HEV in order to be classified as a PHEV by the IEEE [1] .

1) A 4 kWh battery that is used for propulsion must be installed in the vehicle.

2) A connection to the electric grid must be present to recharge the battery.

3) A vehicle must be able to drive 10 or more miles solely using electric power.

Each PHEV used for the analysis presented in this paper meets these requirements with a battery capacity of 16.5 kWh and all electric range of 38 miles [2]. AC PHEV chargers are typically broken up into three categories. These categories are listed in [3] and summarized in Table 1. Based on the charging stations recommended for the PHEV used in this study by [4], only Levels 1 and 2 chargers are considered. A shorter charge time for the PHEV used in this study is listed when a Level 2 charger is used [2]. It is assumed that most consumers will choose the faster charge time if available, so a maximum charge/discharge rate is chosen to correspond to a Level 2 charger.

One of the services that PHEVs may be able to provide to an electric utility is peak shaving through vehicle to grid (V2G). In V2G, PHEV batteries are charged during off peak energy usage times and discharged during peak energy usage. One method of implementing this type of scheme involves using time-of-use (TOU) rates. TOU rates charge more for energy used during peak usage periods than for energy used during off peak usage periods [5] . Using TOU rates in a V2G scheme, a PHEV owner would charge the vehicle battery during off peak pricing periods and discharge the battery back to the electric grid during peak pricing periods.

V2G schemes typically require a communication network to be installed to allow utilities to communicate with PHEV chargers. Also, adding PHEVs utilizing V2G to a radial distribution system that typically sees only unidirectional power flows may create bidirectional power flows on the system. Both the additional communication network and bidirectional power flows lead to a high cost associated with V2G [6] . Furthermore, PHEVs are only capable of supplying a small amount of power in comparison to the overall system load of a large utility. Due to these factors, using PHEVs for peak shaving through V2G will likely face many challenges. Some existing studies attempt to conduct an economic analysis of V2G, however they either do not consider actual vehicle driving cycle data, fail to include the costs of negative impacts on the battery, or are conducted on large scale systems where the implementation of V2G is expected to be slow to gain traction [6] - [8] . Also, studies fail to show the breakeven point for PHEV owners in terms of TOU rates.

This research aims to implement V2G on a university campus system or other large commercial campus, henceforth referred to as V2C. This type of system will alleviate some of these concerns while still allowing the potential for both the consumer and utility to benefit. In the system being studied, which is the Clemson Univer- sity distribution system, the campus is seen as a large net consumer by the electric utility, similar to a large commercial building. Therefore, a large number of vehicles can be connected before all of the campus loads would be supplied by sources other than the electric utility. This would eliminate the issue of two way power flows on the electric utility system caused by the campus or other large commercial building supplying power back to the electric utility.

Also, by using TOU rates, a decentralized method of control is used to command ve- hicle charging and discharging in order to greatly reduce or eliminate the need for communication. The electric utility also benefits through a reduction of load power drawn by the university campus during peak energy usage times. The economic analysis on such a scheme is conducted to show at what point it becomes economically beneficial for PHEV owners to participate. An economic comparison with stationary battery energy storage is also conducted.

The structure of this paper is as follows: Section 2 contains details of spatial and temporal PHEV models used for this study; Section 3 contains specifics of the vehicle to campus algorithm; Section 4 has details and the results for the economic analysis; Section 5 shows the peak shaving impacts of the vehicle to campus algorithm; and Section 6 contains the final conclusion of this paper. Notation may be found at the end of the paper.

Table 1. PEV charging levels [3] .

Charging Level	Supply Voltage	Maximum Current	Real Power
1	120 V, 1 ph	12 A	1.44 kW
2	208 - 240 V, 1 ph	32 A	6.66 - 7.68 kW
3	208 - 600 V, 3 ph	400 A	>7.68 kW

2. VEHICLE SPATIAL AND TEMPORAL DISTRIBUTIONS

In order to complete an accurate analysis of V2C, the locations of PHEVs during different times of the day are needed. Both temporal and spatial distributions of PHEVs must be determined. The spatial distribution is used to determine how far a PHEV has traveled when it arrives on campus from home and how far it must travel to return home. The temporal distribution is used to determine what time PHEVs arrive on campus and what time they leave to return home. In order to develop accurate distributions, data from the 2009 National Household Travel Survey is used along with the distribution fitting command in MATLAB

[9]. For the spatial distribution, the responses to how far a worker's job is from home are used. These responses correspond to the variable DISTTOWK in [9] . Based on the responses, an exponential distribution with the probability density function shown in Figure 1 is chosen for the distance driven between work and home. The average distance driven between work and home is determined to be 14.1 miles.

The times PHEVs arrive on campus as well as when they leave must also be approximated. For the temporal distribution of when vehicles arrive on campus, the responses to the end time of all trips with the destination of work are used. For the temporal distribution of when vehicles leave campus to return home, the responses corresponding to the start time of all trips with the destination of home are used. These responses correspond to the variables ENDTIME, STRTTIME, and WHYTO for the arrival time, leaving time, and purpose of the trip, respectively. Based on the responses, normal distributions with the probability density functions shown in Figure 2 are chosen for the arrival and departure times. The average arrival time of PHEVs on campus is determined to be 9:07 am and the average departure time of PHEVs from campus is determined to be 3:23 pm.

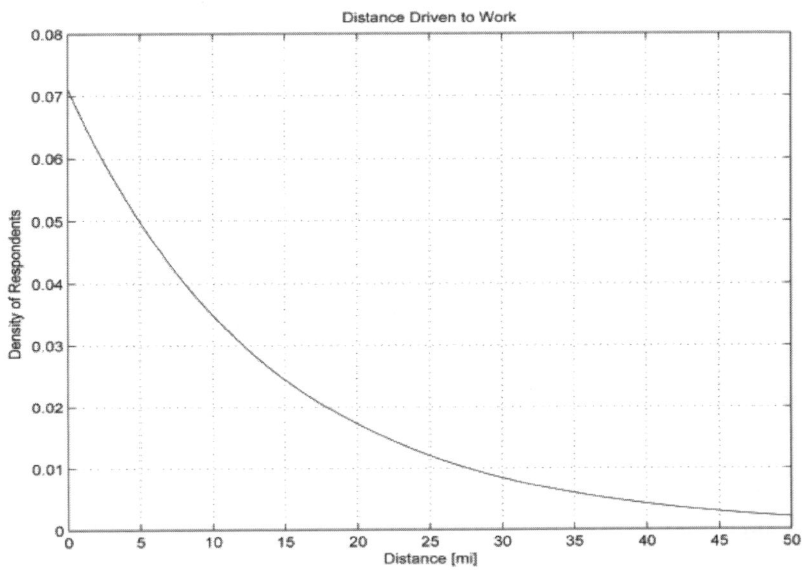

Figure 1. Spatial probability density function.

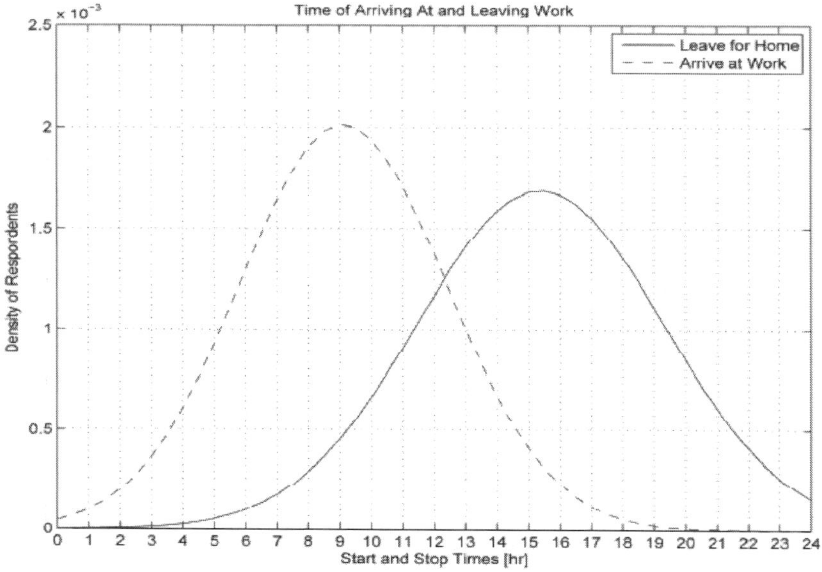

Figure 2. Temporal probability density function.

3. VEHICLE TO CAMPUS ALGORITHM

In order to implement V2C, a decentralized algorithm is implemented. The algorithm requires the PHEV owner to estimate how far the vehicle will be driven after departure from campus and also input the time the PHEV will depart. The goal of the algorithm is to utilize V2C to supply as much energy as possible during the peak pricing period while also maintaining enough charge upon departure that the distance input by the PHEV owner can be traveled solely on electric power. The necessary state of charge (SOC) is calculated using the estimated distance input by the user as well as the 98 MPGe and 16.5 kWh battery capacity given in [2] . Any charging necessary is performed during off peak pricing if possible. The only time on peak charging would occur is the case where a vehicle is not connected long enough during off peak times to reach the necessary SOC required by the PHEV owner.

The algorithm attempts to charge the PHEV as much as possible before the on peak period, limited by a maximum charging rate of 7 kW, which is chosen based on a level 2 charger [3] . The algorithm also attempts to discharge the PHEV as much as possible during on peak periods, limited by maintaining enough energy in the battery at the time of departure and a maximum discharging rate of 7 kW. If the PHEV remains connected after the peak period ends, off peak charging is also allowed during this time. An SOC operating range of 10% - 89.2% is chosen based on 98 MPGe, 38 mile all electric range, and 16.5 kWh battery capacity [2] . An efficiency of 90% is assumed for both charging and discharging operations [8] . Charging and discharging rates are reduced from 7 kW wherever possible such that the rate is the lowest that will allow completion of the energy transfer in the allotted time.

For the economic analysis, it is assumed that the PHEV is driven the average of 14.1 miles to campus and arrives at 9:07 am. It is also assumed that the vehicle will be driven the average of 14.1 miles from campus to home upon departure at 3:23 pm. On peak pricing is assumed to start at 2:00 pm and to end at 8:00 pm [10] - [16] . Based on these parameters, the SOC profile of the PHEV battery while the vehicle is parked in campus is shown in Figure 3. The blue line in Figure 3 represents the energy stored in the PHEV battery over time, starting at the arrival time and ending at the departure time. The vertical red lines show the beginning and end of the on peak pricing period.

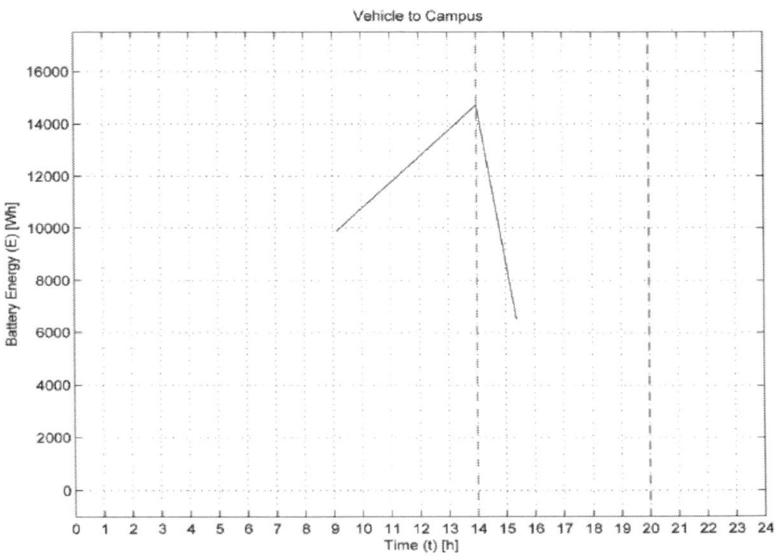

Figure 3. Energy profile of the PHEV battery throughout the day while parked on campus.

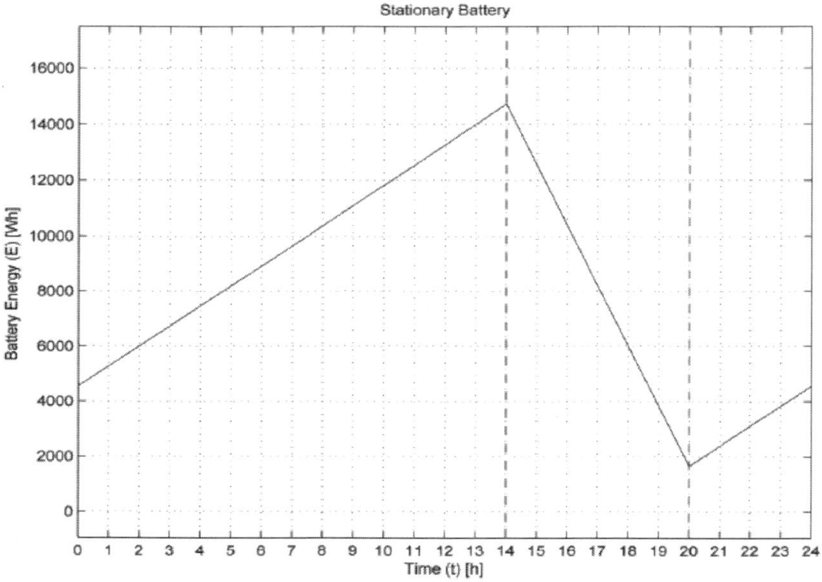

Figure 4. Energy profile of the stationary battery throughout the day.

4. ECONOMIC ANALYSIS

In order to fully examine the economic analysis of a V2C scheme, a comparison with a stationary battery that has the same capacity as the PHEV battery is conducted. The same SOC operating range and efficiency for the PHEV battery are also assumed for the stationary battery. Figure 4 shows the energy stored in the stationary battery throughout the day.

To produce a meaningful comparison for V2C operation, a baseline cost for operating a PHEV without V2C must be considered. For the baseline, the cost of charging the PHEV only during off peak is considered. Also, only the energy necessary to drive to work and make the return trip home is necessary to consider as no other energy is depleted from the PHEV battery. Charging efficiency is also included in the calculation. The baseline cost is calculated using Equation (1). For operation without V2C, there is no accelerated degradation of the battery, so no additional cost to compensate for that is necessary.

$$Cost_{BL} = (E_W + E_H) * \text{Rate}_{Off} \quad (1)$$

Whenever V2C is employed, an accelerated degradation of the PHEV battery occurs compared to without V2C due to the increased cycling of the battery. After repeated cycling, a battery loses some of its usable storage capacity. For this study, a cycle life of 2500 cycles was chosen based on [2] [8]. This must be accounted for in order to give PHEV owners an accurate estimate of whether or not it is economical to participate in V2C. This cost is accounted for using Equation (2). This compensates the PHEV owner for a partial cycling of the battery based on how much energy is discharged from the battery for V2C operations [8] .

$$Cost_C = \frac{Cost_B}{Cyc_R} * \frac{E_D}{E_B} \quad (2)$$

It is stated in [17] that the goal for battery price in PHEVs varies from \$150 - \$500 per kWh. For this study, three different battery costs are studied including the two extremes and one halfway between these estimates. These costs include \$150 per kWh, which corresponds to a battery cost of \$2475, \$325 per kWh, which corresponds to a battery cost of \$5362.50, and \$500 per kWh, which corresponds to a battery cost of \$8250.

The total cost of energy exchange during V2C is given by Equation (3). The total savings that are seen from using V2C are calculated using Equation (4).

$$Cost_E = E_C * Rate_{Off} - E_D * Rate_{On} \quad (3)$$

$$Sav_{V2C} = Cost_{BL} - \left(Cost_C + Cost_E\right) \quad (4)$$

The results are shown in Figure 5 for the three battery costs. The axes show the off peak price, the price difference between peak and off peak prices, and the total savings. In Figure 5(a), Figure 5(c), and Figure 5(e) any points above the black surface represent an economic benefit from utilizing V2C while those below it represent a loss of money through utilizing V2C. Figure 5(b), Figure 5(d), andFigure 5(f) give an overhead view of these surfaces, where the magenta line represents the intersection with the black surface. All points above the line represent an economic benefit from utilizing V2C while those below it represent a loss of money through utilizing V2C.

It can be seen that as battery costs increase, the difference between peak and off peak price must increase in order to realize an economic benefit from V2C. In all cases however, an economic benefit is possible if the difference between peak price and off peak price is large enough. It can also be seen that at the high off peak prices, the difference between peak and off peak prices must be larger than the low off peak prices in order to see an economic benefit. This is due to the charging and discharging efficiencies, both of which are below 100%. A PHEV owner may have to pay for energy losses, in which case this must be included in the analysis for accuracy. At high off peak prices, the cost of energy losses is higher than during low off peak prices.

A similar economic analysis is conducted for a stationary battery. Figure 4 shows a stationary battery's SOC profile and the cycle compensation is again given by Equation (2). In this case however, ED represents the full usable battery capacity, EB. For the stationary battery, the savings are calculated using Equation (5).

$$Sav_{SB} = -\left(Cost_C + Cost_E\right) \quad (5)$$

The results of the economic analysis for the stationary battery can be found in Figure 6. Again, all points above the black surface or magenta line represent an economic benefit from utilizing stationary storage while those below represent a loss of money. It can be seen that the break even line for the PHEV battery and stationary battery is the same. This is due to using the same battery parameters for both comparisons. However, the amount of savings is different between the two analyses. This is due to the different charging and discharging profiles of the PHEV and stationary batteries.

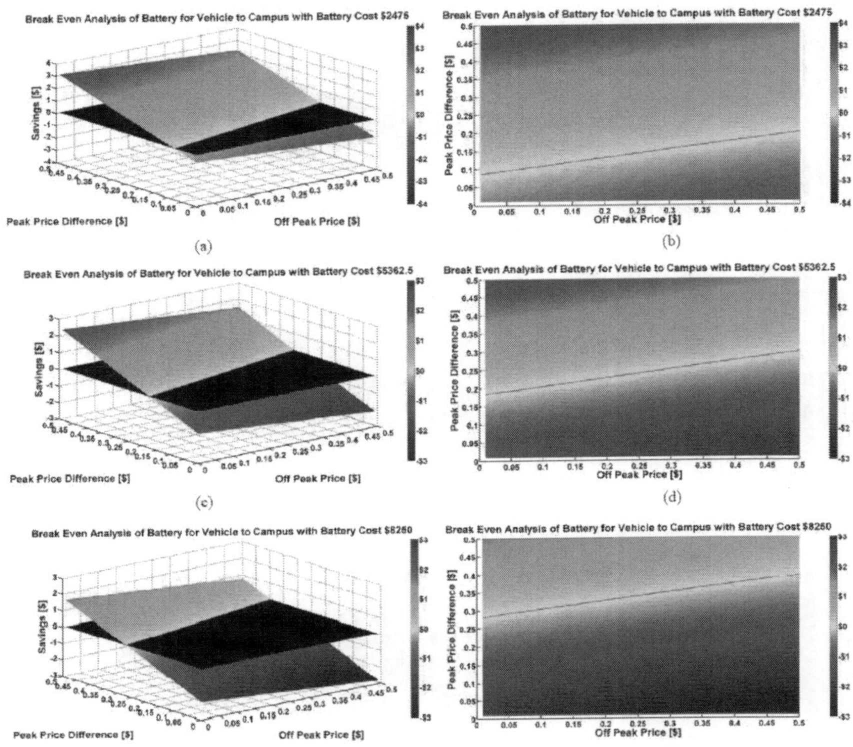

Figure 5. Break even analysis of vehicle to campus using the average cycle. (a) Battery cost—$2475; (b) Battery cost—$2475—overhead view; (c) Battery cost—$5362.50; (d) Battery cost—$5362.50—overhead view; (e) Battery cost—$8250; (f) Battery cost— $8250—overhead view.

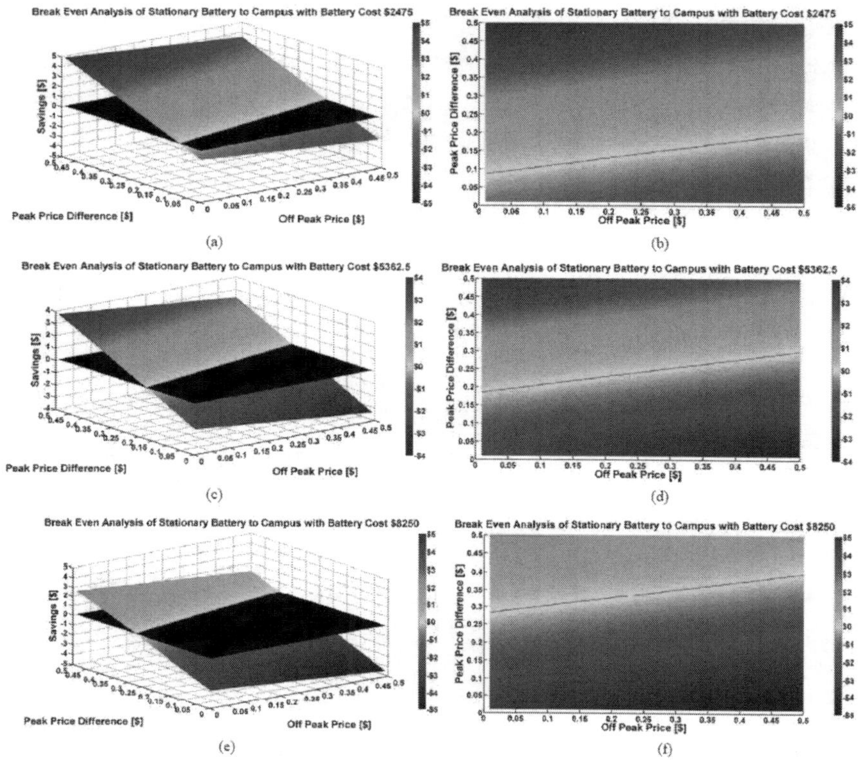

Figure 6. Break even analysis of stationary battery to campus. (a) Battery cost—$2475; (b) Battery cost—$2475—over- head view; (c) Battery cost—$5362.50; (d) Battery cost—$5362.50—overhead view; (e) Battery cost —$8250; (f) Battery cost—$8250—overhead view.

5. PEAK SHAVING IMPACTS

Based on the ability of V2C to prove profitable, it has been shown that V2C has the potential to benefit consumers without impacting required driving behavior. This section focuses on showing the benefit of V2C to the electric utility feeding campus. The V2C algorithm previously described works by charging PHEVs during off peak rate times and discharging them during peak rate times. The peak rate times correspond to the peak load values seen from campus loads. Figures 7-9 show the campus load profile on the 12.47 kV circuit for a spring, summer, and winter day based on a base case simulation, respectively. These load profiles do not include V2C or stationary battery energy storage. It can be seen that the peak load for the spring day occurs between 2 and 3 pm. For the summer day, the peak load also occurs between 2 and 3 pm. The winter day's peak occurs between 12 and 1 pm. It can also be seen that the highest peak load occurs during the spring season.

In order to examine the impacts of adding PHEVs to the campus system while utilizing V2C, a total of 300 PHEVs are added in a total of 8 car parks around campus. The car parks are located based on currently available parking lots on campus. Six of the car parks contain thirty PHEV charging stations each, while the other two contain sixty PHEV charging stations each. The distance driven to campus, arrival time, and departure times are all chosen using random variables corresponding to the distributions described previously, with the constraint that departure time must be later than arrival time. Arrival and departure times are rounded to the nearest minute. Based on these parameters, the V2C algorithm is applied to each vehicle to determine the charge and discharge rates. In order to examine the maximum effects of peak shaving due to PHEVs, it is assumed for this section that the difference between on peak and off peak rates is high enough that V2C proves profitable. The charge and discharge rates from the V2C algorithm are then applied to the 300 PHEVs. Due to the arrival and departure times, not all PHEVs are present on campus during the peak rate period, meaning not all PHEVs participate in peak shaving. Based on the temporal distributions however, it can be seen that the majority of PHEVs that park on campus regularly will be present during the beginning of the peak rate period, with the availability decreasing as the time gets later.

Figures 10-12 show the campus 12.47 kV system load with PHEVs participating in V2C for a spring, summer, and winter day, respectively. It can be seen that the peak system load is reduced for both the spring and summer cases. For the winter case, the load is reduced during the utility peak pricing period, however this does not overlap with the system peak load, so the peak is not reduced. Due to the use of PHEVs in a V2C campus scenario, the electric utility providing electric power to campus can expect a significant reduction of campus load during the peak pricing periods. By drawing less energy during peak rate periods, the campus will also save money on the purchase of electricity.

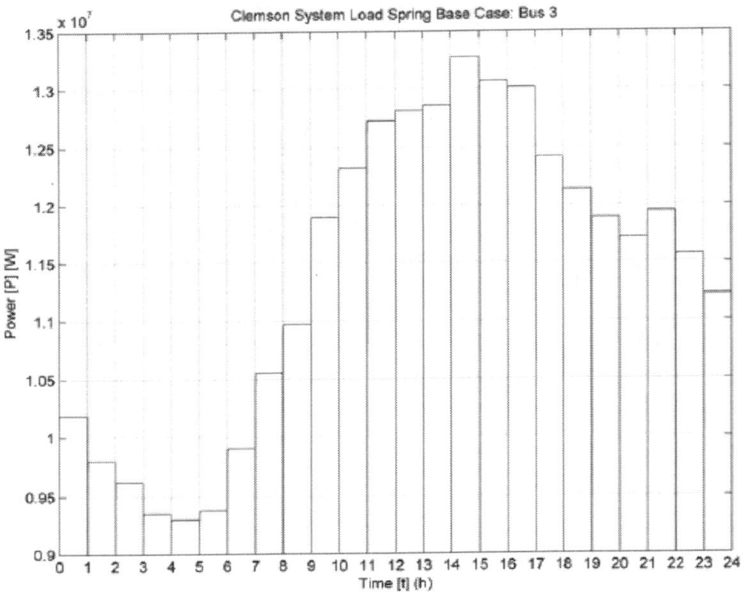

Figure 7. Clemson University 12.47 kV system load—spring- base case.

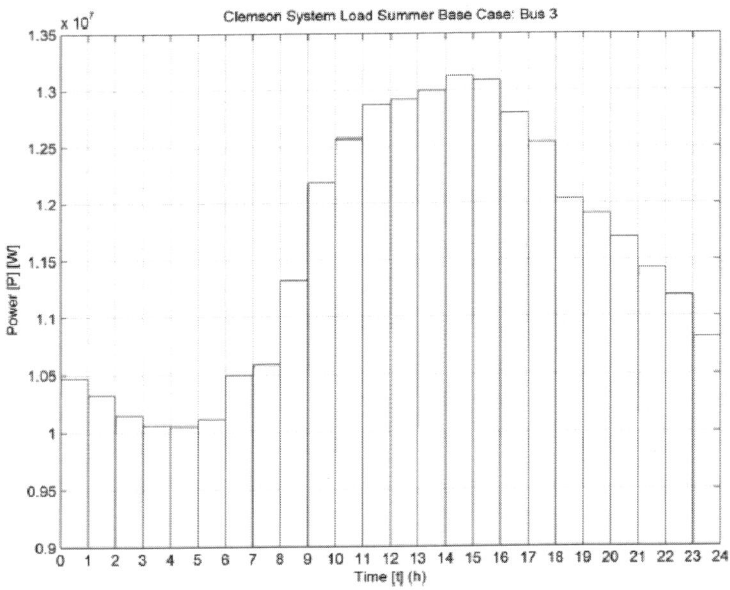

Figure 8. Clemson University 12.47 kV system load—sum- mer-base case.

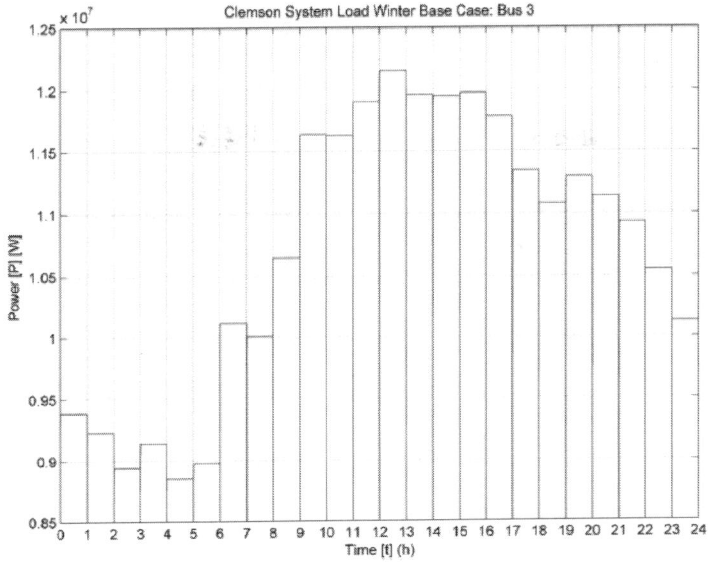

Figure 9. Clemson University 12.47 kV system load—winter-base case.

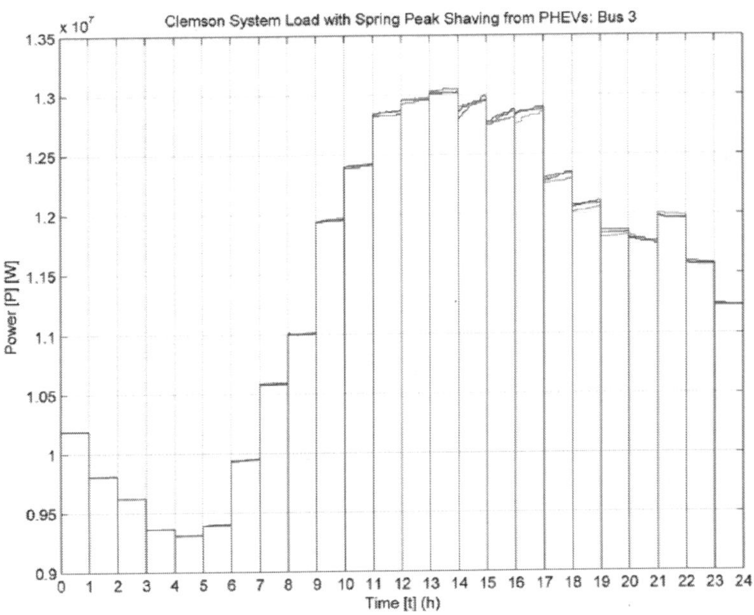

Figure 10. Clemson University 12.47 kV system load—spring-V2C.

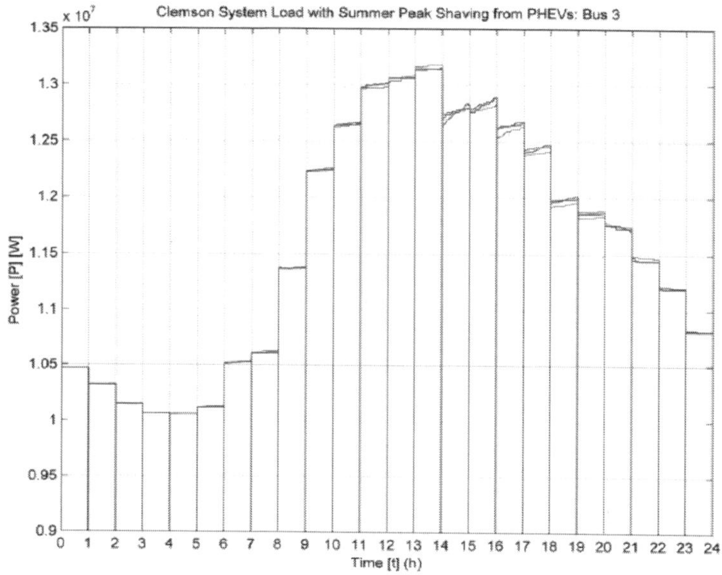

Figure 11. Clemson University 12.47 kV system load—summer-V2C.

In order to show the viability of using V2C as an acceptable source for peak shaving, a comparison with stationary battery energy storage is again conducted. In order to draw a meaningful comparison, each PHEV is replaced with an equivalent sized stationary battery of the same capacity at the same location. The profile of the stationary battery energy is shown in Figure 4. The load profile for the campus with stationary energy storage is shown in Figures 13-15 for a spring, summer, and winter day, respectively.

It can be seen that the peak shaving impacts of the stationary battery energy storage are similar to that of the PHEVs using V2C. The main difference is that more peak shaving occurs with the stationary battery energy storage than with PHEVs using V2C. This is especially apparent during the latter portion of the peak rate time. This is due to the disconnection of PHEVs based on the temporal distributions and the minimum SOC requirement placed on individual PHEVs by the spatial distribution.

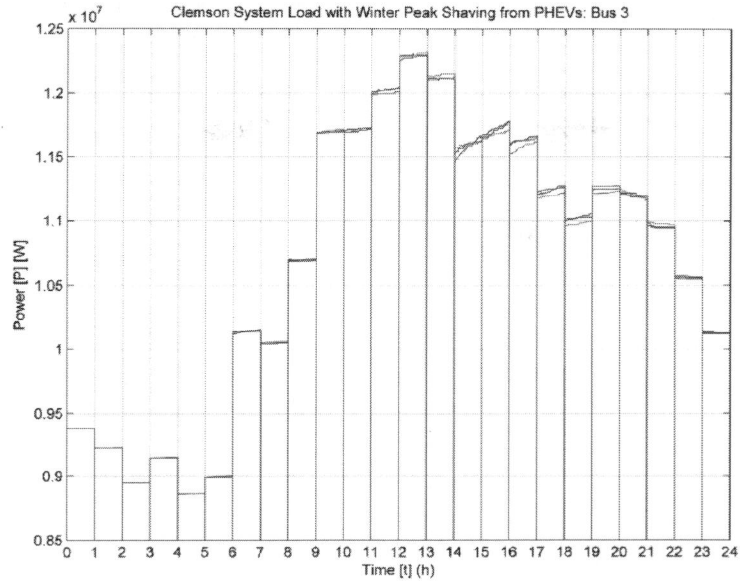

Figure 12. Clemson University 12.47 kV system load—winter-V2C.

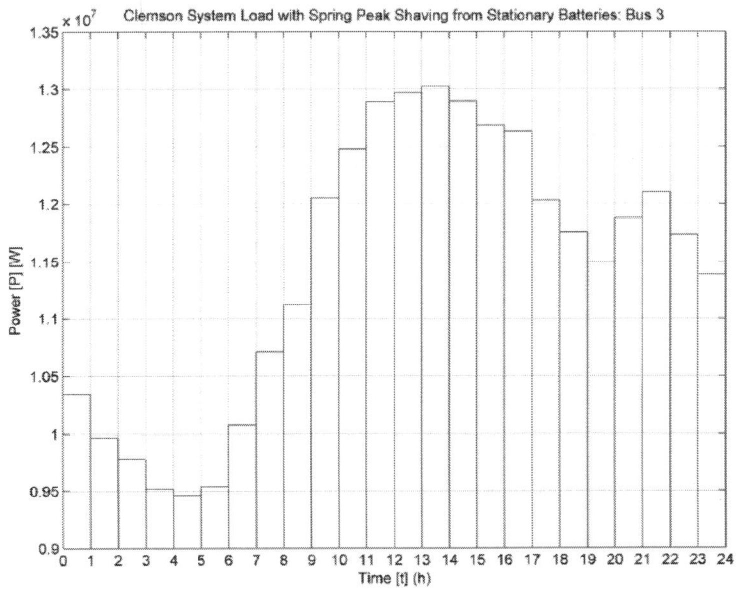

Figure 13. Clemson University 12.47 kV system load—spring-stationary battery.

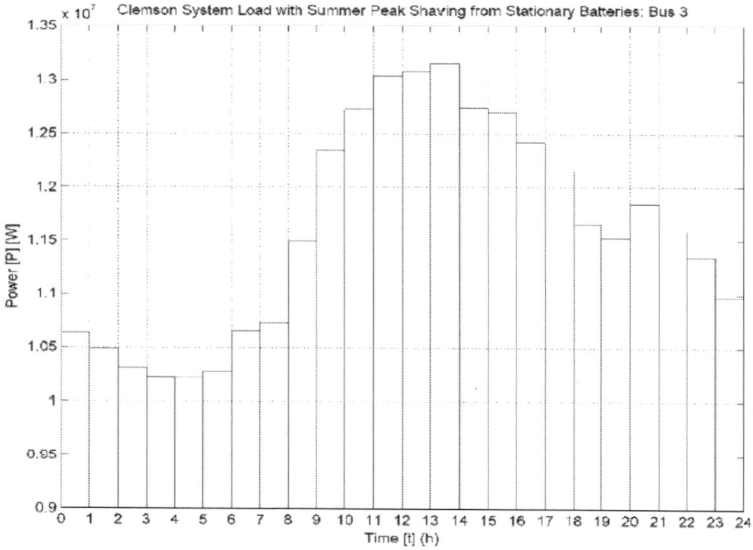

Figure 14. Clemson University 12.47 kV system load—summer-stationary battery.

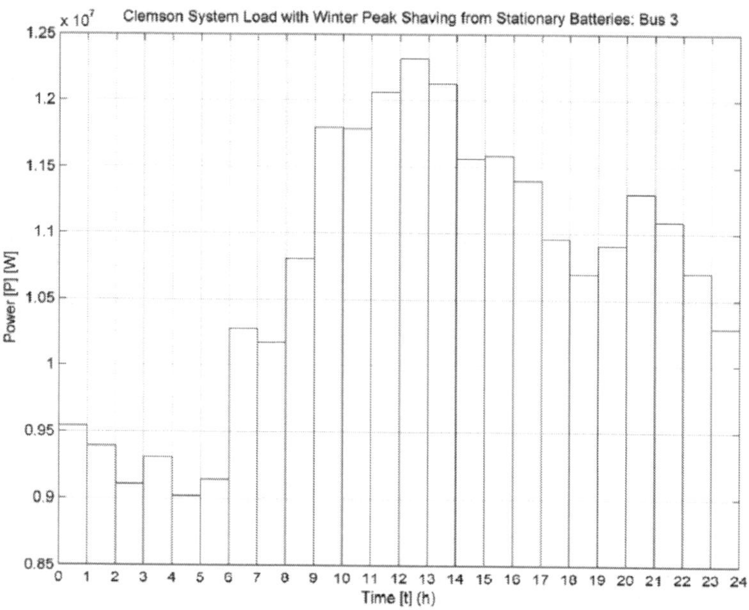

Figure 15. Clemson University 12.47 kV system load—winter-stationary battery.

6. CONCLUSION

An economic analysis and peak shaving analysis for both V2C and stationary storage are conducted. It is shown that V2C and stationary storage can both be cost effective and allow for peak shaving on a university campus. The cost effectiveness and the amount of savings are dependent on the battery cost and electricity pricing during both on and off peak times. Under the V2C scenario, the PHEV owner is left with enough energy to travel a desired distance solely under electric power while still participating in V2C. Implementation of V2C reduces the load of the university campus, as seen by the electric utility, during peak usage times, thus benefiting the electric utility without the concern of two way power flows occurring on the utility system. While a stationary battery has the potential to save more than a PHEV battery, the break even point between the two is the same. Thus, utilizing V2C will allow a university campus to achieve similar benefits without the initial high cost and space requirements of a stationary battery.

ACKNOWLEDGEMENTS

The authors would like to thank the members of the Clemson University Electric Power Research Association (CUEPRA) for their financial support and valuable discussion.

REFERENCES

1 Anumolu, P., Banhazl, G., Hilgeman, T. and Pirich, R. (2008) Plug-In Hybrid Vehicles: An Overview and Performance Analysis. 2008 IEEE Long Island Systems, Applications and Technology Conference, Farmingdale, 2 May 2008, 1-4.

2 Chevrolet (2014) 2014 Volt: Electric Car—Hybrid Car Chevrolet Web.http://www.chevrolet.com/volt-electric-car.html

3 Kisacikoglu, M.C., Ozpineci, B. and Tolbert, L.M. (2010) Examination of a PHEV Bidirectional Charger System for V2G

Reactive Power Compensation. 2010 25th Annual IEEE Applied Power Electronics Conference and Exposition (APEC), Palm Springs, 21-25 February 2010, 458-465.

4 Bosch. (2014) Charging Stations Bosch Electric Vehicle Solutions Web.http:/www.pluginnow.com/charging_stations

5 Davis, B.M. and Bradley, T.H. (2012) The Efficacy of Electric Vehicle Time-of-Use Rates in Guiding Plug-In Hybrid Electric Vehicle Charging Behavior. IEEE Transactions on Smart Grid, 3, 1679-1686.

6 Yilmaz, M. and Krein, P.T. (2013) Review of the Impact of Vehicle-to-Grid Technologies on Distribution Systems and Utility Interfaces. IEEE Transactions on Power Electronics, 28, 5673-5689.

7 Pang, C., Dutta, P. and Kezunovic, M. (2012) BEVs/PHEVs as Dispersed Energy Storage for V2B Uses in the Smart Grid. IEEE Transactions on Smart Grid, 3, 473-482.

8 Das, R., Thirugnanam, K., Kumar, P., Lavudiya, R. and Singh, M. (2014) Mathematical Modeling for Economic Evaluation of Electric Vehicle to Smart Grid Interaction. IEEE Transactions on Smart Grid, 5, 712-721.

9 US Department of Transportation, Federal Highway Administratio (2014) 2009 National Household Travel Survey. http://nhts.ornl.gov

10 PG & E (2014) Time-of-Use.http://www.pge.com/en/ mybusiness/ rates/tvp/toupricing.page?WT.mc_id=Vanity_tou

11 PPL Electric Utilities (2014) Time-of-Use Rate Option. https:// www.pplelectric.com/at-your-service/electric-rates-and-rules/time-of-use-option.aspx

12 Southern California Edison (2014) Residential Rate Plans. https://www.sce.com/wps/portal/home/residential/rates/residential-plan/tou/

13 Portland General Electric (2014) Time of Use: Pricing. http:// www.portlandgeneral.com/residential/your_account/billing_payment /time_of_use/pricing.aspx

14 Con Edison (2014) Voluntary Time-of-Use.http://www.coned. com/ customercentral/energyresvoluntary.asp

15 NV Energy (2014) Residential Time of Use for Southern Service Territory. Time of Use Rate for Home. https://www. nvenergy. com/home/paymentbilling/timeofuse.cfm

16 BGE (2014) Time of Use Pricing.http://www.bge.com/ waystosave/ manageyourusage/pages/time-of-use-pricing.aspx

17 Anderson, D. (2009) An Evaluation of Current and Future Costs for Lithium-Ion Batteries for Use in Electrified Ve- hicle Powertrains. Master's Thesis, Environmental Management, Duke University, Durham.

CHAPTER 2

Multiobjective Optimal Design of an Inverter Fed Axial Flux Permanent Magnet In-Wheel Motor for Electric Vehicles

Christophe Versèle[1], Olivier Deblecker[1] and Jacques Lobry[1]

[1] Electrical Engineering Department, University of Mons, Belgium

1. INTRODUCTION

Battery electric vehicles (EVs) and hybrid EVs are now at the forefront of transportation researches in order to reduce the emission of pollutant gasses, especially in the inner cities. However, the present performances of EVs are far from being competitive to those of traditional combustion engine vehicles, mainly in terms of autonomy. Therefore, power-saving and mass-saving are of the upmost importance when designing the entire drive system of EVs and their electric motor drives (Tseng & Chen, 1997).

In conventional EVs, the power system consists of battery, electric motors with drives, transmission gears and differentials to the wheels (Yang & Chuang, 2007). The mechanical transmission system contributes greatly to the power loss, through the heat dissipation in the various components, and to the weight of the vehicle. An interesting alternative to this conventional power system is the concept of in-wheel motors or hub-in motors as illustrated in Fig.1. In this concept, the motor is directly integrated into the wheel, thus eliminating transmission gears and differentials with their associated power loss (Yang et al., 2004). Furthermore, the elimination of mechanical components in transmission chains or gears reduces the weight of the vehicle. Each of the in-wheel motors has its own voltage supply inverter (VSI) as well as its own speed or torque controller. All the in-wheelmotors are coordinated by a digital vehicular-speed controller with differential gear.

Due to their robustness, low cost, performances and simplicity of design, induction motors (IMs) are often preferred for EVs propulsion. However, in recent years,Neodymium-Iron-Bore (NdFeB) axial flux permanent magnet (AFPM) motors have become an interesting alternative to IMs due to their compactness, low weight and high torque density. Moreover, AFPM motors are pancake-type, fit perfectly the wheel of an automobile vehicle and, thus, can be easily and compactly integrated into the wheel. According to these properties, axial flux motors seem to be a better choice than more conventional radial flux motors for this kind of application. Note that this issue is discussed in (Versèle et al., 2009).

There are many alternatives for the design of AFPM motors (Sahin & Vandenput, 1999): slotted or slotless stator, rotor with interior or surface-mounted permanent magnets (PMs), internal or external rotor, numbers of rotors and stators, etc. In this chapter, a double-sided motor with internal slotted stator and surface-mounted PMs is proposed as basic design choice essentially motivated by the presence of two air gaps doubling the torque.

Among the various researches, many authors have published papers about the optimal design of in-wheel motors for EVs or wheelchairs (Espanet et al., 1999; Nilssen et al., 2005; Tseng & Chen 1997;Yang et al., 2004) as well as about the optimization of AFPM motors (Azzouzi et al., 2006; Chun et al., 2007; Cvetkovski & Petkovska, 2002).

In this chapter, the objective is to optimize simultaneously one of the four in-wheel motors of an EV and its own VSI in terms of weight and power loss. Note that the simultaneous optimization of the in-wheel motor and its own VSI, rarely discussed in literature, results in a system optimized towards the requirements of the EV.

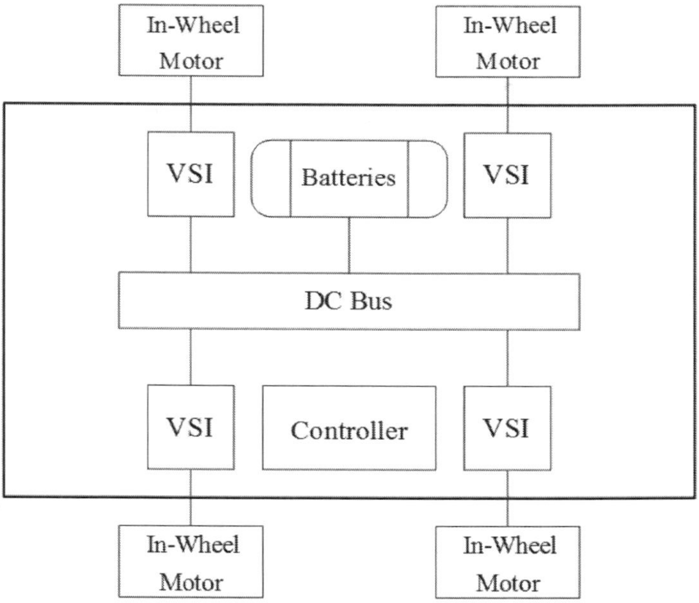

Figure 1. In-wheel motors drive

To do so, a multiobjective optimization (MO) technique based on evolutionary algorithms (EAs) is used. EAs are stochastic search techniques that mimic natural evolutionary principles to perform search and optimization procedure. Among the several approaches to evolutionary optimization, GAs have been chosen and the so-called Elitist Nondominated Sorting Genetic Algorithm (NSGA-II) (Deb, 2002) is used to perform the optimal design of the in-wheel motor and its own VSI. Note that GAs are chosen because they have already proved their efficiency to optimize every kind of electrical machines (Skaar & Nielssen, 2003) and power electronics converters (Helali, 2005; Malyna et al., 2007; Versèle et al., 2010).

The remainder of this chapter is organized as follows. First, the requirements in terms of power and torque of an EV are discussed in Section 2. Then, Section 3 describes the AFPM motor and VSI models used in the design procedure. In Section 4, the MO technique based on GAs as well as the proposed optimization routine are described. Finally, a design example is exposed in Section 5 and the advantages and limitations of the new design procedure are discussed in Section 6.

2. REQUIREMENTS OF AN EV

In order to determine the requirements, in terms of power and torque, of one of the four in-wheel motors (considering an EV driven by four in-wheel motors), a computer model of an EV traction system is presented in this section.

The road load on the vehicle consists of three forces (Yang & Chuang, 2007): (1) the rolling resistance F_r, (2) the aerodynamic drag force F_a and (3) the climbing force F_c which are expressed as (Ehsani et al., 2005):

$$F_r = f_r M_v g \tag{1}$$

$$F_a = 0.5 \rho A_f C_D (v + v_w)^2 \tag{2}$$

$$F_c = M_v g \sin(\alpha) \tag{3}$$

where f_r is the rolling resistance coefficient (which is an empirical coefficient depending on the road-tire friction), M_v is the mass of the vehicle, g is the earth gravity acceleration, ρ is the air density, A_f is the frontal area of the vehicle, C_D is the coefficient of aerodynamic resistance (that characterizes the shape of the vehicle), v is the vehicle speed, v_w is the component of the wind speed on the vehicle's moving direction and α is the road angle (deduced from the road slope). According to Newton's second law, the total tractive effort F_t required to reach the desired acceleration a and to overcome the road load is:

$$F_t = M_v a + F_r + F_a + F_c \,. \tag{4}$$

Once the total tractive effort is computed, the total torque $T\,t$ and power $P\,t$ required to be produced by the four in-wheel motors can be expressed as:

$$T_t = F_t r_{wheel} \tag{5}$$

$$P_t = T_t \Omega_{wheel} \tag{6}$$

where r_{wheel} is the drive wheels radius and Ω_{wheel} is the rotational wheels speed.

Based on the specifications of an urban EV (Ehsani et al., 2005), summarized in Table 1, and on the above-described EV traction system, the requirements of one in-wheel motor can be easily computed. All the results are presented in Table 2. Note that, in addition to provide its requirements, the in-wheel motor must also respect some constraints. The main constraints are the total weight of each of the four in-wheel motors $M\,motor$ (imposed by the maximal authorized "unsprung" wheel weight) and the imposed outer radius $R\,out$ of the motor (imposed by the rim of the wheel). Those are also specified inTable I and Table 2.

Table 1. Specifications of an EV

Weight M_v	1150 kg
Max. speed v_{max}	13.9 m/s (50 km/h)
Acceleration a	1 m/s²
Frontal area A_f	2.5 m²
Coefficient of aerodynamic resistance C_D	0.32
Rolling resistance coefficient f_r	0.015
Max. road angle α	5.7° (10%)
Rim diameter	14"
Number of in-wheel motors	4

Table 2. Requirements of one in-wheel motor

Torque T_t	> 107 Nm
Power P_t	> 8.7 kW
Weight of the motor M_{motor}	< 43.125 kg

3. MODELING OF THE AFPM MOTOR AND VSI

In order to evaluate the two objective functions, viz. the weight and the losses of the motor and the VSI, and to verify if the constraints are not violated during the design procedure, two models are necessary: one for the motor and one for the VSI. It should be noticed that analytical models have been chosen in this paper with the aim of reducing the computational time. These models permit to evaluate the weight and the losses of the motor and the VSI as well as to estimate the torque and power developed by the motor.

3.1. Afpm Motor Model

Analytical design of AFPM motors is usually performed on the average radius R_{ave} of the machine (Parviainen et al., 2003) defined by:

$$R_{ave} = (R_{in} + R_{out})/2 \tag{7}$$

where R_{in} and R_{out} are respectively the inner and outer radius of the machine (see Fig. 2(a)).

The use of the average radius as a design parameter allows evaluating motor parameters and performances based on analytical design methods (Gieras et al., 2004).

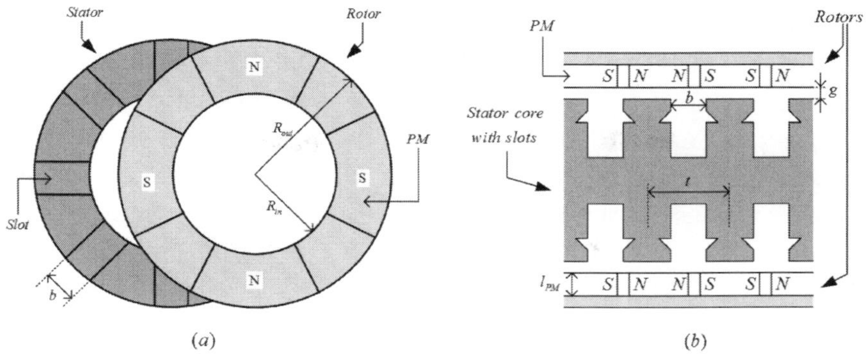

Figure 2. a) Stator and rotor of an AFPM machine and (b) doubled-sided AFPM machine with internal slotted stator

The air gap flux density B_g is calculated using the remanence flux density B_r (in the order of 1.2 T for a NdFeB type PM) and the relative permeability μ_{ra} of the PM as well as the geometrical dimensions of the air gap and the PM (thickness and area) according to:

$$B_g = k_{\sigma PM} \frac{B_r}{1 + \mu_{ra} \dfrac{k_C g}{l_{PM}} \dfrac{S_{PM}}{S_g}} \tag{8}$$

where g and l_{PM} are respectively the air gap thickness and PM thickness (see Fig. 2(b)); S_g and S_{PM} are respectively the air gap area and PM area. Finally, in (8), $k_{\sigma PM}$ (<1) is a factor that takes into account the leakage flux and k_C (>1) is the well-known Carter coefficient.

On the one hand, in order to obtain an accurate estimation of the air gap flux density and torque developed by the motor, the factor $k_{\sigma PM}$ is one of the most essential quantities that must be computed. Indeed, the leakage flux has a substantial effect on the flux density within the air gap and PMs (Qu & Lipo, 2002) and, therefore, on the torque developed by the motor (see (11)). In addition to air gap leakage flux, zigzag leakage flux is another main part of the leakage flux. The zigzag leakage flux is the sum of three portions (Qu & Lipo, 2002): the first part of the zigzag leakage flux is short-circuited by one stator tooth, the second part links only part of the windings of a phase and the third part travelling from tooth to tooth does not link any coil. Note that, in this paper, an analytical model developed by Qu and Lipo (Qu & Lipo, 2002) for the purpose of the design of surface-mounted PM machines is used to compute the factor $k_{\sigma PM}$. This model permits to express this factor in terms of the magnetic material properties and dimensions of the machine. It is thus very useful during the design stage.

On the other hand, the main magnetic flux density in the air gap decreases under each slot opening due to the increase in reluctance. The Carter coefficient permits to take into account this change in magnetic flux density caused by slot openings defining a fictitious air gap greater than the physical one. It can be computed as follows (Gieras et al., 2004):

$$k_C = \frac{t}{t - \gamma g} \tag{9}$$

where t is the average slot pitch (see Fig. 2(b))and γ is defined by:

$$\gamma = \frac{4}{\pi}\left[\frac{b}{2g}\arctan\left(\frac{b}{2g}\right) - \ln\sqrt{1+\left(\frac{b}{2g}\right)^2}\right] \tag{10}$$

where b is the width of slot opening (see Fig. 2(b)).

Assuming sinusoidal waveform for the air gap flux density and the phase current, the average electromagnetic torque T of a double-sided AFPM motor can be calculated by:

$$T = 2\pi B_g A_{in} R_{out}^3 (k_d - k_d^3) \tag{11}$$

where A_{in} is the linear current density on the inner radius of the machine and k_d is the ratio between inner and outer radii of the rotor disk. It should be noticed that, for a given outer radius and magnetic and electric loading, the factor k_d is very important to determine the maximum torque developed by the motor. So, this factor will be one of the optimization variables. Figure 3 reports the per-unit (p.u.)

electromagnetic torque with respect to this factor k_d. One can remark that the maximum value of the torque is reached for $k_d \approx 0.58$.

The electromagnetic power P can easily be calculated by the product of torque and rotational speed Ω_r of the motor according to:

$$P = T\Omega_r \,. \tag{12}$$

The double-sided AFPM motor losses are the sum of the stator winding losses, the stator and rotor cores losses, the PMs losses and the mechanical losses, whereas its weight is the sum of the stator and the two rotors weights, the stator winding weight and the PMs weight. Note that the computation of those different parts of the two objective functions can easily be found elsewhere (Gieras et al., 2004) and, so, it is not described in this chapter.

Finally, it should also be pointed out that some electrical parameters of the AFPM motor, such as the stator resistance (R_s) and the direct (L_d) and quadrature (L_q) axes inductances, can be calculated once the motor has been design using the above-described fundamental design equations.

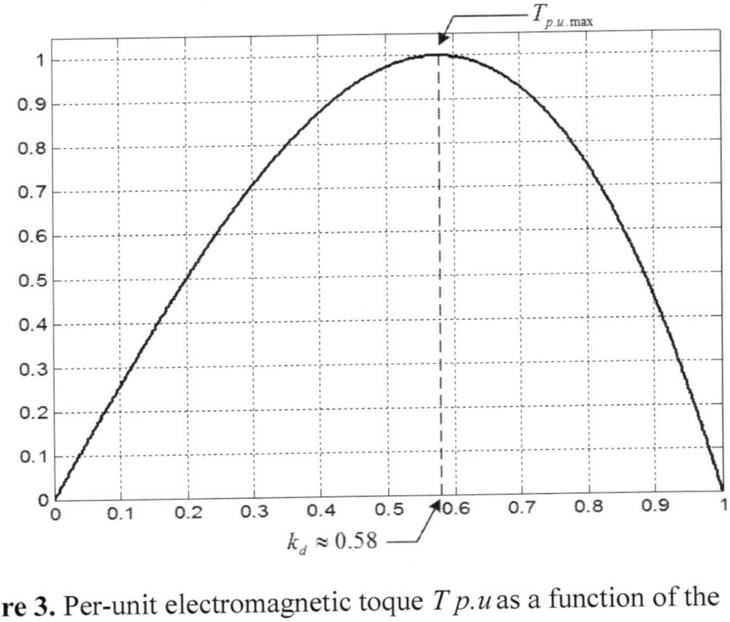

Figure 3. Per-unit electromagnetic toque $T_{p.u}$ as a function of the factor k_d

3.2. Afpm Motor Model Validation

In order to validate the analytical AFPM motor model presented in this chapter, analytical and experimental results are compared. To do so, the proposed model is applied to a 5.5 kW, 4000 rpm AFPM motor. The calculated motor parameters are then compared with parameters obtained by classical tests (test at dc level, no-load test, etc.) performed on an existing AFPM pump motor. All the results are reported in Table 3. As can be seen, very small differences are obtained between the analytical and experimental results, whatever the parameters. According to this validation method, one can conclude that the proposed analytical design process gives reasonable results in this particular case and can be used in the optimization procedure of the in-wheel motor of an EV.

Table 3. Analytical and experimental results for the 5.5 kW, 4000 rpm AFPM pump motor

Parameters	Analytical results	Experimental results
R_s (dc-resistance)	0.45 Ω	0.423 Ω
L_d	0,129 H	0,117 H
L_q	0,127 H	0,117 H
T	13.97 Nm	15.04 Nm
P	5.85 kW	5.5 kW

3.3. Vsi Model

The total loss of the semiconductor devices (IGBTs and diodes) of the VSI (employing sinusoidal pulse-width-modulation) consists of two parts: the on-state losses and the switching losses. The on-state losses of the devices are calculated from their average I_{ave} and rms I_{rms} currents by the well-known expression (Semikron, 2010):

$$P_{on} = V_{th} \cdot I_{ave} + r \cdot I_{rms}^2 \tag{13}$$

where V_{th} represents the threshold voltage and r the on-state resistance, both taken from the manufacturer's data sheets.

The switching losses are, as for them, calculated by the following formula (Semikron, 2010):

$$P_{sw} = f_s \cdot \left(E_{turn-on} + E_{turn-off}\right)_{T_j, I_C, V_{CE}} \tag{14}$$

where $E_{turn\text{-}on}$ and $E_{turn\text{-}off}$ are the energies dissipated during the transitions, both taken from the manufacturer's data sheets, at given junction temperature Tj, on-state collector current IC and blocking voltage VCE. Note that the average and rms values of the current used in (13) can easily be computed.

Based on the total loss of the semiconductor devices, the heatsink can be designed in order to limit the junction temperature to a predefined temperature (typically in the order of125 °C). This temperature can be estimated from the ambient temperature Ta, the thermal resistances (junction-case: $R\,th,jc$, case-heatsink: $R\,th,ch$ and heatsink-ambient: $R\,th,ha$) and the total loss $P\,sc$ of all the semiconductor devices by:

$$T_j = T_a + \left(R_{th,jc} + R_{th,ch} + R_{th,ha}\right) \cdot P_{sc} \,. \tag{15}$$

From (15), the thermal resistance of the heatsink $R\,th,ha$ needed to limit the junction temperature to the predefined value can be computed and, then, the heatsink can be selected from the manufacturer's data sheets.

The total weight of the VSI is the sum of the weight of all the semiconductor devices and the weight of the heatsink.

4. OPTIMIZATION ROUTINE BASED ON THE NSGA-II

As mentioned previously, in this contribution, a MO technique based on EAs is used. Those are stochastic search techniques that mimic natural evolutionary principles to perform the search and optimization procedures (Deb, 2002).

GAs have been chosen because they overcome the traditional search and optimization methods (such as gradient-based methods) in solving engineering design optimization problems (Deb & Goyal, 1997). Indeed, there are, at least, two difficulties in using traditional optimization algorithms to solve such problems. Firstly, each traditional optimization algorithm is specialized to solve a particular type of problems and, therefore, may not be suited to a different type. As this is not the case with the GAs, no particular difficulties have been met to adapt the considered GA (viz. the NSGA-II, see below) to the multiobjective optimal design of the AFPM motor and its VSI. Only the models of these converters had to be used in combination with the GA in order to evaluate the values of the considered objectives. Secondly, most of the traditional methods are designed to work only on continuous variables. However in engineering designs, some variables are restricted to take discrete values only. In this chapter, this requirement arises, e.g., for the choice of the number of poles pairs.

Mixed-variable optimization problems are difficult to tackle because they pose the problems of the combinatorial and continuous optimization problems (Socha, 2008). For this reason, there are not many dedicated algorithms in literature and most of the approaches used in these algorithms relax the constraints of the problem. The most popular approach consists in relaxing the requirements for the discrete variables which are assumed to be continuous during the optimization process (Deb & Goyal, 1997). This type of approach is, often, referred as continuous relaxation approach.

Apart from the relaxation-based approach, there are methods proposed in literature that are able to natively handle mixed-variable optimization problems. However, only a few such methods have been proposed. Among them, the Genetic Adaptive Search is based on the fact that there are versions of the GAs dedicated to discrete variables and other versions dedicated to continuous variables. So, the GAs can be easily extended to natively handling both continuous and discrete variables. Such an approach has been proposed in (Deb & Goyal, 1997) and will be used in this chapter as it has already proved to be efficient to solve engineering problems (see, e.g. (Deb & Goyal, 1997)). Pattern Search Method (Audet & Dennis, 2001), Mixed Bayesian Optimization Algorithms (Ocenasek

& Schwarz, 2002) and Ant Colony optimization (Socha, 2008) are other methods which permit to tacklemixed-variable problems.

Among the several MO techniques using GAs (see, e.g., (Deb, 2007)), the so-called NSGA-II (Deb *et al.*, 2002), described in the next Section, will be used to perform the optimal design.

4.1. Nsga-II

NSGA-II is a recent and efficient multiobjective EA using an elitist approach (Deb, 2002). It relies on two main notions: nondominated ranking and crowding distance. Nondominated ranking is a way to sort individuals in nondominated fronts whereas crowding distance is a parameter that permits to preserve diversity among solutions of the same nondominated front.

The procedure of the NSGA-II is shown in Fig. 4 and is as follows (Deb, 2002). First, a combined population $R\,t$ (of size 2 N) of the parent $P\,t$ and offsprings $Q\,t$ populations (each of size N) is formed. Then, the population $R\,t$ is sorted in nondominated fronts. Now, the solutions belonging to the best nondominated set, i.e. $F\,1$, are of best solutions in the combined population and must be emphasized more than any other solution. If the size of $F\,1$ is smaller than N, all members of $F\,1$ are inserted in the new population $P\,t+1$. Then, the remaining population of $P\,t+1$ is chosen from subsequent nondominated fronts in order of their ranking. Thus, the solutions of $F\,2$ are chosen next, followed by solutions from $F\,3$. However, as shown in Fig. 4, not all the solutions from $F\,3$ can be inserted in population $P\,t+1$. Indeed, the number of empty slots of $P\,t+1$ is smaller than the number of solutions belonging to $F\,3$. In order to choose which ones will be selected, these solutions are sorted according to their crowding distance (in descending order) and, then, the number of best of them needed to fill the empty slots of $P\,t+1$ are inserted in this new population. The created population $P\,t+1$ is then used for selection, crossover and mutation (see below) to create a new population $Q\,t+1$, and so on for the next generations.

NSGA-II has been implemented in Matlab with real and binary coding schemes. So, a discrete variable is coded in a binary string whereas a continuous variable is coded directly. Such coding schemes are used in this paper because the considered optimization variables (see Table 4 in Section 5) belong to the two categories.

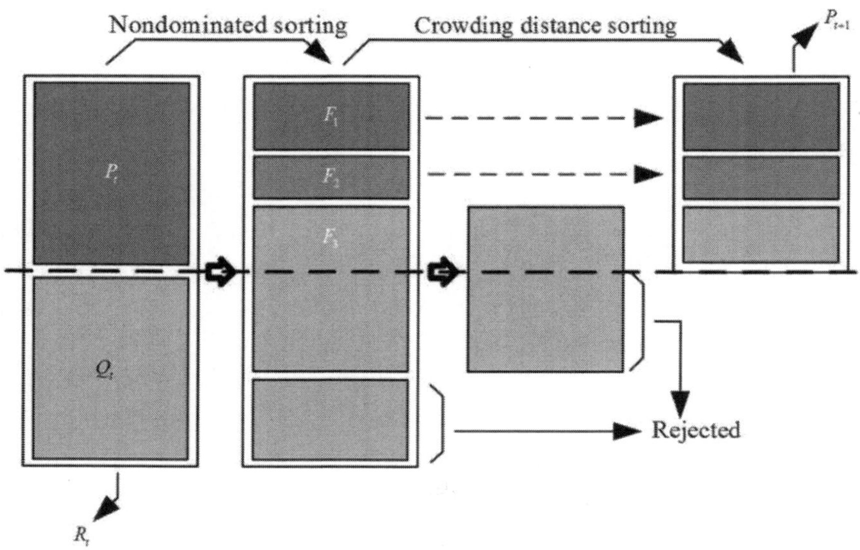

Figure 4. NSGA-II procedure **(Deb, 2002)**

These coding schemes allow a natural way to code different optimization variables, which is not possible with traditional optimization methods. Moreover, the real coded scheme for the continuous variables eliminates the difficulties (Hamming cliff problem and difficulty to achieve arbitrary precision) of coding such variables with a binary scheme.

So, e.g., with the coding scheme used in this paper, the structure of the chromosome (composed by the seven considered optimization variables) of the solution #3 (see Table 5 in Section 5) is as follows:

$$\begin{array}{c} \underbrace{(k_d)(J)(l_{PM})(g)(f_s)}_{\text{real coded variables}} \quad \underbrace{(p)(q)}_{\text{binary coded variables}} \\ \downarrow \\ (0.89)(4.9)(5)(1.9)(1972)\underbrace{(1110)}_{\to 14}\underbrace{(11011110)}_{\to 222} \end{array} \tag{16}$$

There are three fundamental operations used in GAs: selection, crossover and mutation. The primary objective of the selection operator is to make duplicate of good solutions and eliminate bad solutions in a population, in keeping the population size constant. To do so, a tournament selection (Deb, 2002) based on nondominated rank and crowding distance of each individual is used. Then, the selected individuals generate offsprings from crossover and mutation operators. To cross and to mutate the real coded variables the Simulated Binary Crossover and Polynomial Mutation operators (Deb & Goyal, 1997) are used in this chapter. The single-point crossover (Deb, 2002) is, as for it, used to cross the discrete optimization variables. Note that to mutate this type of variables, a random bit of their string is simply changed from '1' to '0' or *vice versa*.

Finally, the constraints must be taken into account. Several ways exist to handle constraints in EAs. The easiest way to take them into account in NSGA-II is to replace thenon-dominated ranking procedure by a constrained non-dominated ranking procedure as suggested by its authors elsewhere (see, e.g., (Deb, 2002)). The effect of using this constrained-domination principle is that any feasible solution has a better nondominated rank than any infeasible solution.

It is important to emphasize that the GA must be properly configured. The size of the population is one of the important parameters of the GA as well as the termination criterion. In this contribution, the size of the population *N* is taken equal to 100. It is important to note that, on the one hand, *N* should be large enough to find out small details of the Pareto front whereas, on the other hand, *N* should not be too large to avoid long time optimization. The termination criterion consists in a pre-defined number of generations which is here also fixed to 400. Finally, the crossover probability and the mutation probability are respectively chosen to be 0.85 and 0.015 as typically suggested in literature (Deb, 2002).

4.2. Design Procedure

The overall design procedure, presented in Fig. 5, has been implemented in the Matlab environment. First, a random initial population is generated. Then, the objective functions, i.e. the total weight and the total losses of the VSI-fed AFPM in-wheel motor, are evaluated based on the initial population and on the above-described models (see Section 3). A convergence test is then performed to check for a termination criterion. If this criterion is not satisfied, the reproduction process using genetic operations starts. A new population is generated and the previous steps are repeated until the termination criterion is satisfied. Otherwise, the Pareto front, i.e. the nondominated solutions within the entire search space, is plotted and the optimization procedure ends.

5. DESIGN EXAMPLE

In order to illustrate the design procedure, a VSI-fed AFPM in-wheel motor with the specifications given in Tables 1 and 2 is designed in this Section.

The lower and upper bounds of the seven considered optimization variables, viz. the factor $k\,d$, the current density in the conductors J, the air gap thickness g, the PMs thickness $l\,PM$, the number of poles pairs p, the number of slots q and the switching frequency $f\,s$, are specified in Table 4. Note that the variables p and q are discrete ones whereas the others are continuous.

It should also be recalled that the in-wheel motor must provide the requirements of the EV as well as respect some constraints. The main constraints are the total weight $M\,motor$ of each of the four in-wheel motor and the imposed outer radius of the motor $R\,out$. Note that these constraints have already been specified in Tables 1 and 2.

The results, i.e. the Pareto front, are presented in Fig. 6. Each point of this Pareto front represents an optimal VSI-fed AFPM in-wheel motor that respects all the constraints. Moreover, the values of the optimization variables corresponding to three particular solutions of the front are detailed in Table 5.

For a practical design, one particular solution of the Pareto front should be chosen. On the one hand, the choice of this particular solution can be let to the designer who can choose *a posteriori* which solution best fits the under consideration application or which objective function to promote. Moreover, in industrial framework, this set of solutions can be confronted with additional criteria or engineer's know-how not included in models.

On the other hand, the designer can also use some dedicated techniques to choose a particular solution of the Pareto front. These can be categorized into two types (Deb, 2002): post-optimal techniques and optimization-level techniques.

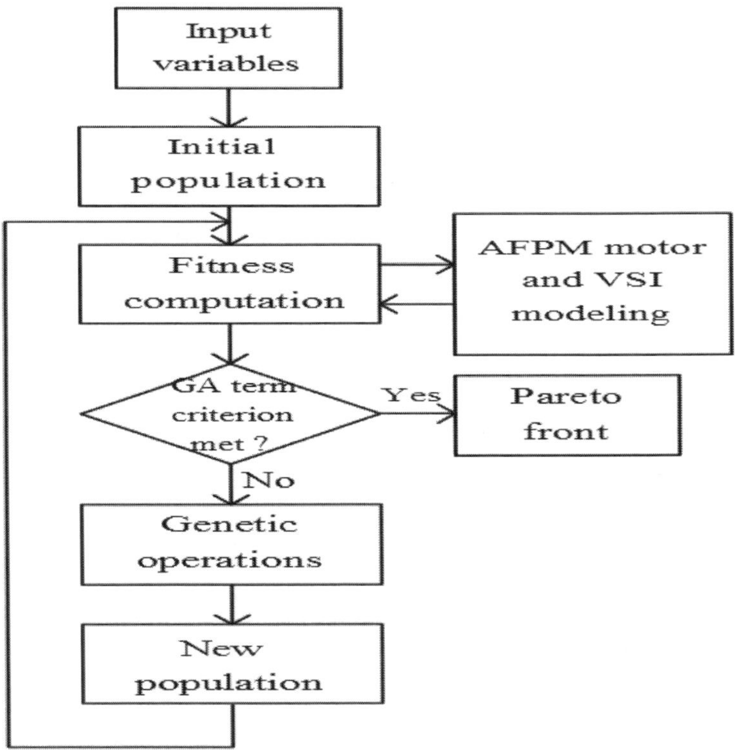

Figure 5. Flowchart of the design procedure using Gas

In the first approach, the solutions obtained from the optimization technique are analyzed to choose a particular solution whereas, in the second approach, the optimization technique is directed towards a preferred region of the Pareto front. Therefore, only the techniques belonging to the first category are helpful in this chapter. Among these techniques, the Compromise Programming Approach (CPA) (Yu, 1973) is often used in multiobjective problems. The CPA picks a solution which is minimally located from a given reference point (e.g. the ideal point which is a nonexistent solution composed with the minimum value of the two objectives). Note that other techniques, such as the Marginal Rate of Substitution Approach (Miettinen, 1999), the Pseudo-Weight Vector Approach (Deb, 2002) or a method based on a sensitivity analysis (Avila *et al.*, 2006), can also be used.

Table 4. Optimization variables

Variables	Bounds	Type
k_d	[0.5 ; 0.9]	Continuous
J	[1 ; 6] A/mm²	Continuous
l_{PM}	[1 ; 20] mm	Continuous
g	[1 ; 5] mm	Continuous
f_s	[1 ; 10] kHz	Continuous
p	[1 ; 15]	Discrete
q	[50 ; 255]	Discrete

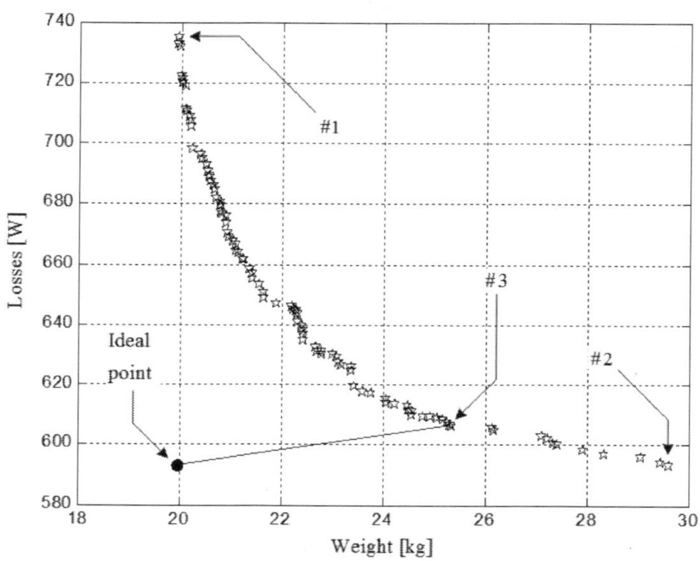

Figure 6. Pareto front

Table 5. Details of three particular solutions

	#1	#2	#3
k_d	0.89	0.89	0.88
J [A/mm^2]	4.9	5.5	4.9
l_{PM} [mm]	5	5	5
g [mm]	1.9	3.2	1
f_s [Hz]	1972	2110	1410
p	14	15	10
q	222	255	158
T_t [Nm]	207	208	207
P_t [kW]	8.72	8.77	8.72
M_{Motor} [kg]	17.2	15.1	26

For instance, using the CPA, the solution #3 of the Pareto front is minimally located from the ideal point. This solution can therefore be considered for a practical design and corresponds, moreover, to a good trade-off between the two objectives.

The evolution of the percentage of individuals belonging to the first nondominated front during the optimization procedure is shown in Fig. 7. From this figure, one can easily conclude that all the individuals are located in the first front at the end of this procedure. Moreover, it can also be observed that new nondominated solutions have been found after approximately 150 generations.

From Fig. 8, it can be conclude that all the individuals respect all the constraints since the fourth generation.

In order to study more in details the evolution of the optimization variables, their values have been plotted along the Pareto front (as a function of the weight) in Fig. 9 to Fig. 15. From these figures, it can be concluded that some of them have converged to an optimal value.

First, the optimal value of the factor $k\,d$ is around 0.90 (see Fig. 9). Based on the Fig. 3, one can expect that this value would have converged to 0.58. However, this result is not surprising. Indeed, it has been show elsewhere (Azzouzi et al., 2006) that the maximum value of the torque to weight ratio is obtained for values of $k\,d$ around 0.85 and, in this chapter, the weight must be minimized.

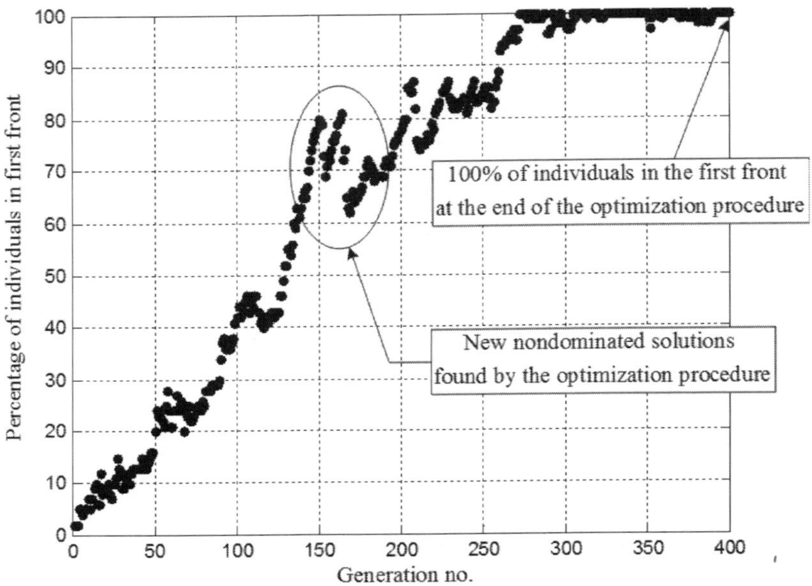

Figure 7. Evolution of the percentage of individuals belonging to the first nondominated front during the optimization procedure

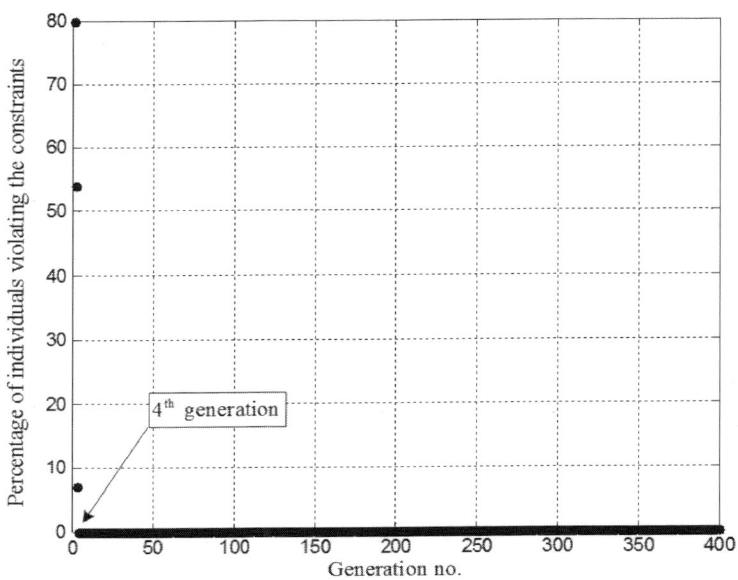

Figure 8. Evolution of the percentage of individuals violating the constraints during the optimization procedure

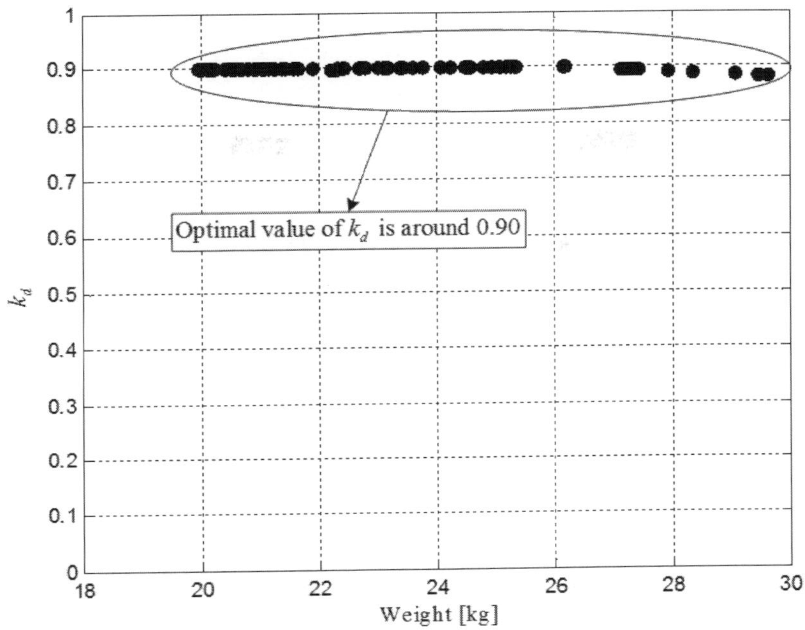

Figure 9. Evolution of *k d* along the Pareto front

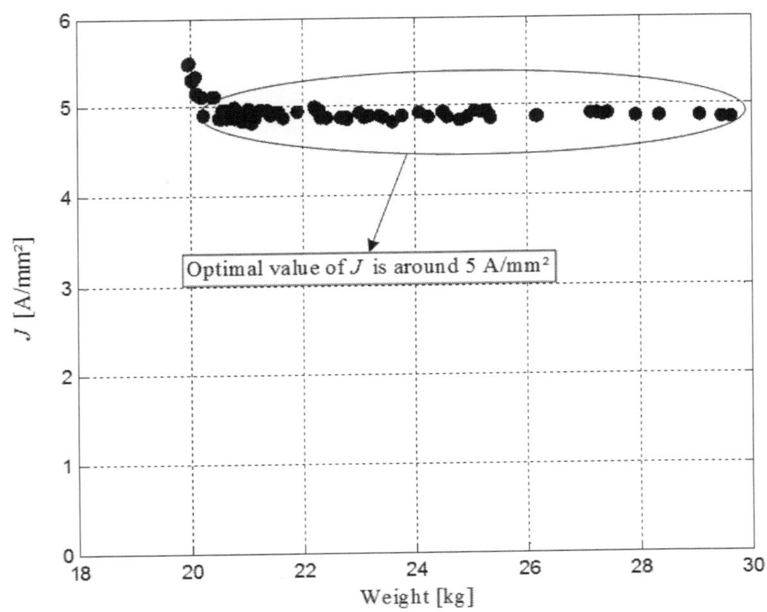

Figure 10. Evolution of *J* along the Pareto front

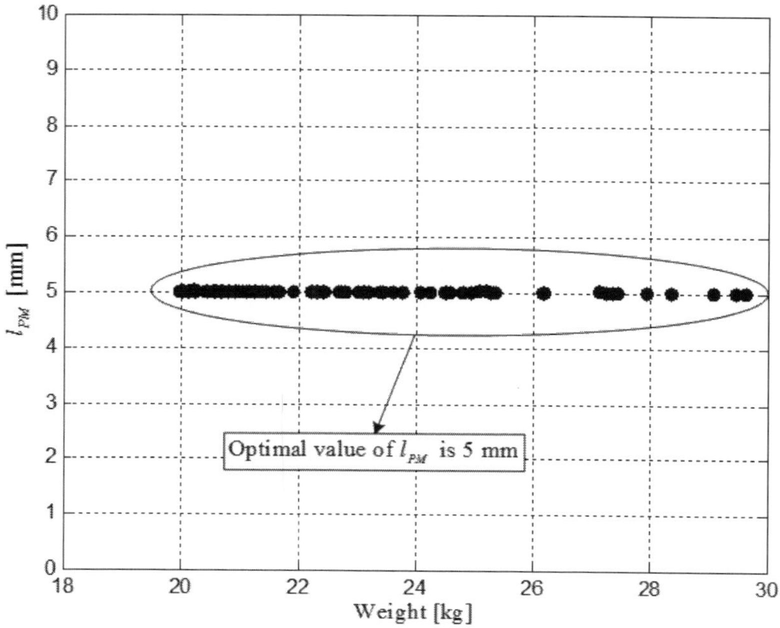

Figure 11. Evolution of *l PM* along the Pareto front

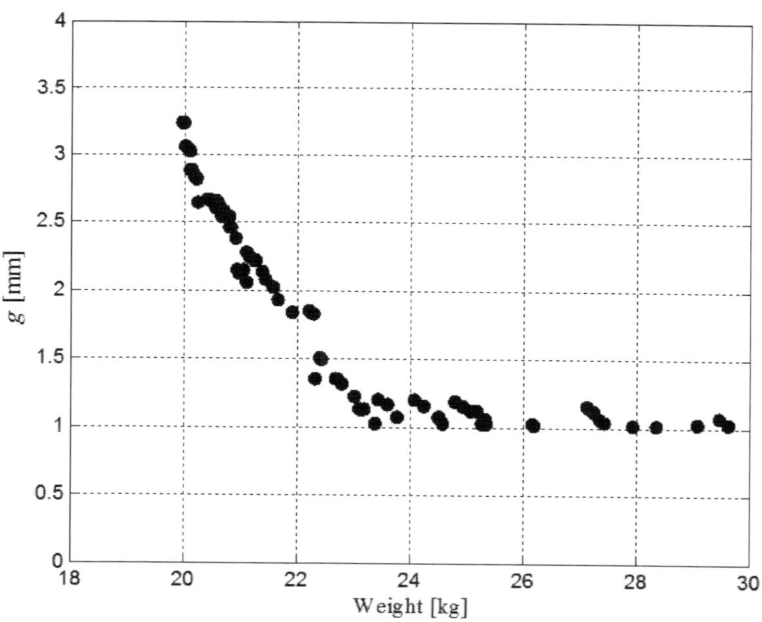

Figure 12. Evolution of *g* along the Pareto front

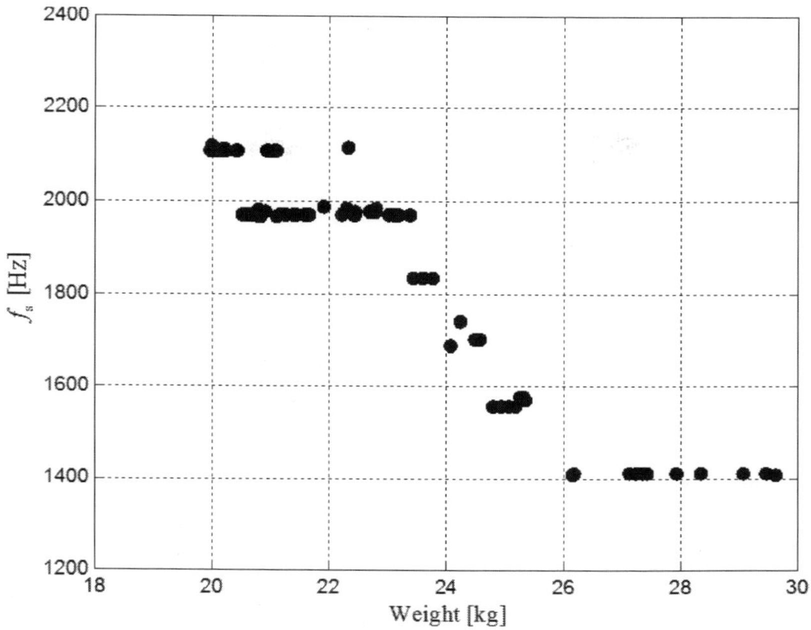

Figure 13. Evolution of *f s* along the Pareto front

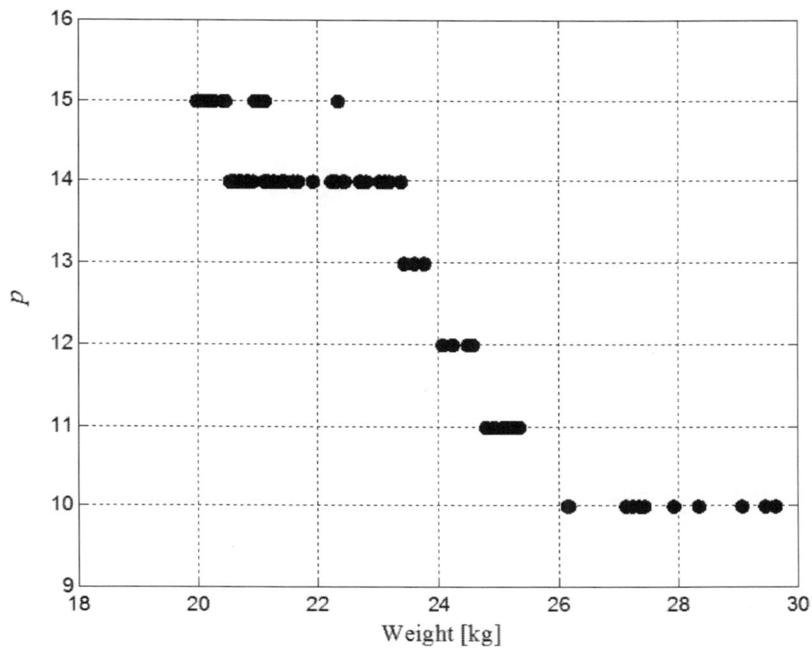

Figure 14. Evolution of *p* along the Pareto front

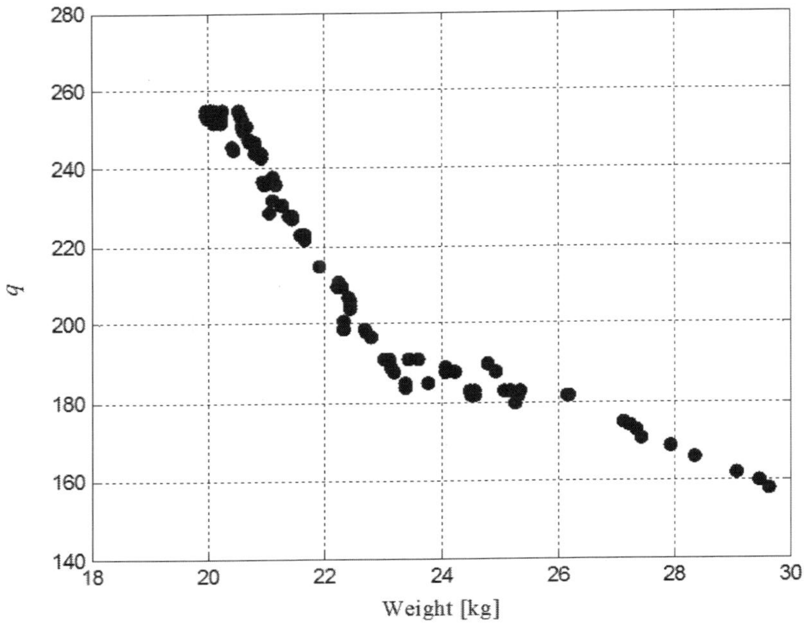

Figure 15. Evolution of q along the Pareto front

Second, the optimal value of J is around 5 A/mm^2 as shown in Fig. 10 and, third, $l\ PM$ has converged to the optimal value of 5 mm (see Fig. 11).

In order to further analyze the optimization results, a study of the correlation level between the optimization variables and the objective functions is performed. The results are graphically represented in Fig. 16 and discussed below. Note that a positive value of the correlation factor indicates that the objective function grows when the optimization variable grows whereas a negative value indicates that the objective function reduces when the optimization variable grows.

From Fig. 16, it can be concluded that the optimization variables which have not converged to an optimal value, viz. g, fs, p and q, have a significant influence on the two objective functions. Indeed, the correlation coefficient between each variable and each objective are, in absolute value, equal or greater than 0.8. The correlations are therefore strong. The fact that the correlation factors between $k\ d$ or J and the objective value are smaller is due to the fact that these variables have converged around an optimal value. Moreover, the fact that $l\,PM$ has converged to an optimal value leads to a correlation coefficient close to zero.

One can easily conclude that the variables p and fs have the bigger influence on the weight (correlation coefficients equal to -0.94) whereas the variable g has the greater influence on the power loss (correlation coefficient equal to 0.98). Figure 16 also justifies the use of a MO technique. Indeed, the two objective functions are conflicting with respect to the optimization variables (except for $l\,PM$) since the correlation levels are of opposite signs.

Finally, the distribution of the weight and the power loss among the AFPM motor and the VSI for the three particular solutions presented in Table 5 are respectively shown in Fig. 17 and Fig. 18 respectively. Figures 19 and 20 present, as for them, the distribution of these two objectives among the several parts of the motor (stator, rotor, PMs and windings).

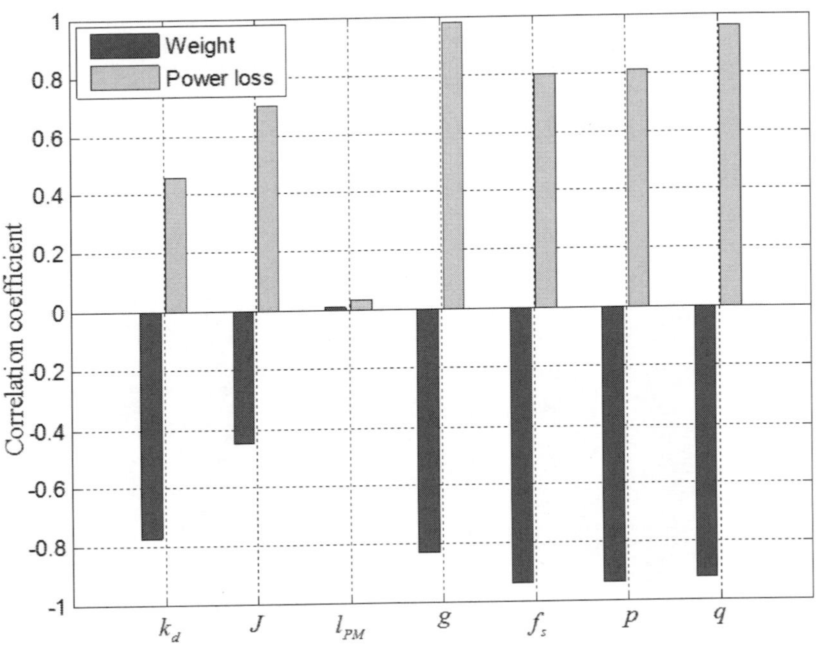

Figure 16. Correlation between the optimization variables and the objective functions

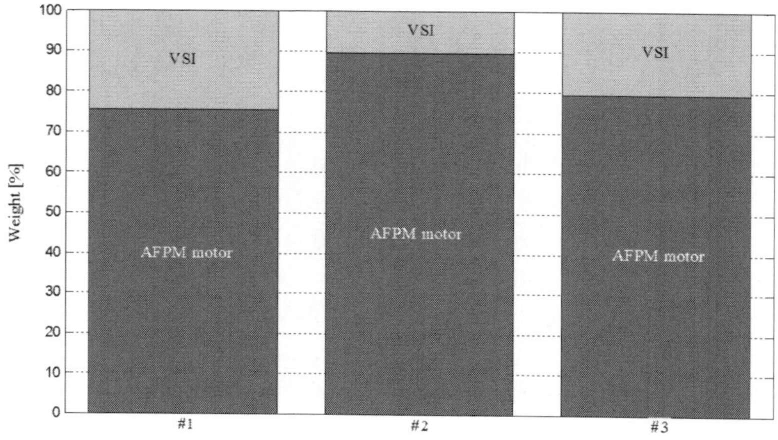

Figure 17. Distribution of the weight among the AFPM motor and the VSI

Figure 17 shows that the major contributor in weight is the AFPM motor. Indeed, it represents at least 75% of the total weight. Instead, the power loss is, as for it, more equally distributed among the two parts. In fact, approximately 60% of the loss is due to themotor (see Fig. 18).

Figure 19 shows that the PMs are the minor contributor in terms of weight and represent more or less 5% to 10% of the total weight whatever the solution. One can also conclude that the rotor is the heavy part of the AFPM motor.

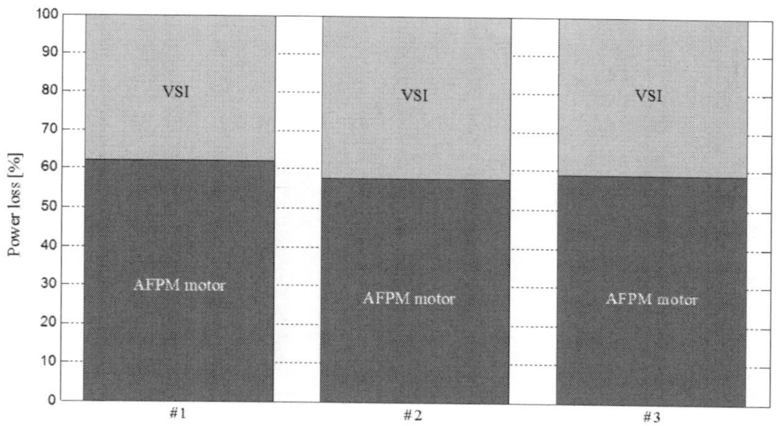

Figure 18. Distribution of the power loss among the AFPM motor and the VSI

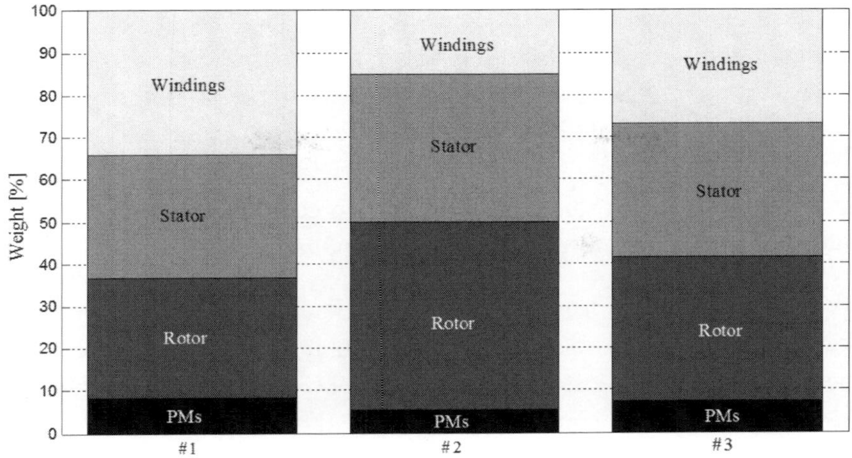

Figure 19. Distribution of the weight among the several parts of the AFPM motor

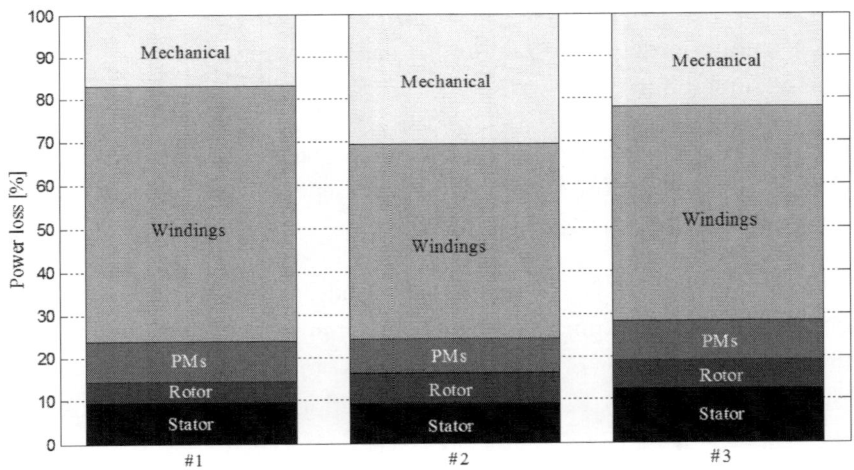

Figure 20. Distribution of the power loss among the several parts of the AFPM motor

In solution #2, the stator and rotor weights largely dominate the winding and PMs weights compared with the other solutions. This can be explained regarding the value of the flux density in the air gap. Indeed, this value is the greatest among the three solutions (in the order of 0.90 T). So, the stator and rotor must be thick enough to avoid saturation of steel. As these thicknesses are computed using the air gap flux density, the stator and rotor of the solution #2 are thicker and, therefore, heavier. Moreover, the windings of the solution #2 are lighter than the windings of the two other solutions since the frequency is greater. Indeed, according to Faraday's law (Mohan et al., 2003), the windings turns necessary to obtain a same electromotive force is less than in the other solutions which, therefore, yields lighter windings realization.

As can be easily seen from Fig. 20, the mechanical and, more especially, the winding losses are the major contributors in terms of power loss. The fact that the winding losses are so high can be explained by the medium frequency effects (skin and proximity effects).

6. ADVANTAGES AND LIMITATIONS OF THE DESIGN PROCEDURE

The design procedure proposed in this paper presents several advantages but also some limitations.

A first advantage of this design procedure is that it is multiobjective. So, several conflicting objectives, often present in engineering design problems, can be optimized simultaneously.

A second advantage is the number of solutions considered in a small time. Indeed, the optimization procedure compares a large number of solutions (in the order of several thousands) to retain only the best in a time of approximately 120 s (for 400 generations of 100 individuals with a Pentium (R) D CPU 3.40 GHz, 3 Go RAM).

A third advantage is the simultaneous optimization of the in-wheel motor and its own VSI. Indeed, it results in a system optimized towards the requirements of the EV.

At last, the design procedure has the major advantage that a set of optimalsolutions – instead of a single one – is proposed to the designer who can choose *a posteriori* which objective function to promote and, then, select a particular VSI-fed in-wheel motor. So, a degree of freedom is still available at the end of the optimization procedure. Moreover, in industrial framework, this set of solutions can be confronted with additional criteria or engineer's know-how not included in models.

The main limitation of the design procedure is related to the analytical modeling of the AFPM motor. Indeed, an analytical modeling of such motor can lead to a lack of accuracy in some cases, e.g. with a PM shape of higher complexity when the magnet occupation ratio varies along the radius of the rotor (Parviaien et al., 2003). This lack of accuracy arises from the reduction of the 3D design problem to a 2D design problem performed on the average radius of the machine.

It should also be noticed that the most accurate method to predict the performances of an AFPM motor is a 3D finite element analysis (FEA) but it is often too much time consuming to be included in an optimal design procedure in which a large numbers of solutions have to be evaluated. Therefore, the optimal design procedure presented in this contribution is very useful during the first stages of the design, although more sophisticated methods, such as 2D or 3D FEA, are required in more advanced phases of the design.

Note that, in this chapter, the magnet occupation ratio has been considered as constant along the radius of the rotor and, therefore, this limitation is not taken into account in the design of the AFPM motor.

7. CONCLUSION

This chapter has addressed the problem of MO design of a VSI-fed AFPM synchronous motor using GAs and a new design procedure has been proposed. The weight and the losses of both the motor and the VSI have been chosen as objective functions whereas the factor $k\,d$, the current density in the conductors, the air gap thickness, the PMs thickness, the number of poles pairs, the number of slots and the switching frequency have been chosen as optimization variables. Finally, the design procedure has been illustrated by the design of a VSI-fed AFPM in-wheel motor for an urban EV and some conclusions have been drawn.

Finally, recall that, although many authors have published papers about the optimal design of in-wheel motors for EVs, the simultaneous optimization of the in-wheel motor and its own VSI has, to the authors' knowledge, rarely been discussed in literature.

REFERENCES

1. C. Audet, Dennis Jr, J. , 2001Pattern search algorithms for mixed variable programming. SIAM Journal on Optimization, 11 3July 2001), 573 594 1052-7189

2. S. L. Avila, A. C. Lisboa, L. Krähenbühl, W. P. Carpes Jr, J. A. Vasconcelos, R. R. Saldanha, R. H. C. Takahashi, 2006Sensitivity Analysis Applied to Decision Making in Multiobjective Evolutionary Optimization, IEEE Transactions on Magnetics, 42 4April 2006), 1103 1106 0018-9464

3. J. Azzouzi, R.. Belfkira, N. Abdel-Karim, G. Barakat, B. Dakyo, 2006Design Optimization of an Axial Flux PM Synchronous Machine: Comparison Between DIRECT Method and GAs Method, Proceedings of 12th International Power Electronics and Motion Control Conference, 1-42440-121-6Slovenia,September 2006.

4. Y. D. Chun, D. Koo, H. , Y. Cho, H. , 2007Multiobjective optimization design of axial flux permanent magnet motor. International Journal of Applied Electromagnetics and Mechanics, 25 1January 2007), 613 619 0827-3284

5. G. Cvetkovski, L. Petkovska, 2008Efficiency Maximisation in Structural Design Optimisation of Permanent Magnet Synchronous Motor, Proceedings of 18th International Conference on Electrical Machines, 978-1-42441-736-0Vilamoura, Portugal, September 2008.

6. K. Deb, M. Goyal, 1997Optimization Engineering Designs Using a Combined Search, Proceedings of 7th International Conference on Genetic Algorithms, 1-55860-299-2Lansing, USA, July 1997.

7. K. Deb, 2002 Multi-Objective Optimization using Evolutionary Algorithms, John Wiley & Sons, Inc., 0-47074-361-1Jersey, USA.

8. K. Deb, 2007Current trends in evolutionary multi-objective optimization, International Journal for Simulation and Multidisciplinary Design Optimization, 1 1December 2007), 1 8 1779-6288

9. M. Ehsani, Y. Gao, S. E. Gay, A. Emadi, 2006 Modern Electric, Hybrid Electric and Fuel Cell Vehicles: Fundamentals, Theory, and Design, CRC Press, 0-84933-154-4USA.

10. C. Espanet, J. M. Kauffmann, F. Wurtz, J. Bigeon, 1999Application of a new optimization approach to the design of electrical wheels. IEEE Transactions on Energy Conversion, 14 4December 1999), 952 958 0013-4457

11. J. F. Gieras, R. Wang, J. , M. J. Kamper, 2004 Axial Flux Permanent Magnet Brushless Machines, Kluwer Academic Publishers, 978-1-40206-993-2Netherlands.

12. H. Helali, et al. 2005Power converter's optimisation and design. Discrete cost function with genetic based algorithms, Proceedings of 11th Power Electronics and Applications Conference, 9-17636-385-6Germany, September 2005.

13. D. V. Malyna, J. L. Duarte, M. A. M. Hendrix, F. B. M. van Horck, 2007Optimization of Combined Thermal and Electrical Behavior of Power Converters Using Multi-Objective Genetic Algorithms, Proceedings of 12th Power Electronics and Applications Conference, 0-78039-457-7Denmark, September 2007.

14. K. Miettinen, 1999 Nonlinear Multiobjective Optimization, Kluwer Academic Publishers, 978-0-79238-278-2Boston, USA.

15. N. Mohan, T. Undeland, W. Robbins, 2003 Power Electronics: Converters, Applications and Design, John Wiley & Sons, 978-0-47122-693-2Hoboken, USA.

16. R. Nilssen, S. E. Skaar, R. Lund, T. Skjellnes, S. Øvrebø, E. Løvli, 2005Design of a permanent magnet synchronous integrated in the wheel rim on wheelchairs, Proceedings of 11th Power Electronics and Applications Conference,9-17636-385-6Germany, September 2005.

17. J. Ocenasek, J. Schwarz, 2002Estimation distribution algorithm for mixed continuous-discrete optimization problems, Proceedings 2nd Euro-International Symposium on Computational Intelligence, 1-58603-256-9 Nederland,September 2002.

18. A. Parviaien, M. Niemelä, J. Pyrhönen, 2003Analytical, 2D FEM and 3D FEM Modeling of PM Axial Flux Machine, Proceedings of 10th Power Electronics and Applications Conference, 9-07581-506-9France, September 2003.

19. R. Qu, T. A. Lipo, 2002Analysis and Modeling of Airgap & Zigzag Leakage Fluxes in a Surface-Mouted-PM Machine, Proceedings of 37th Industry Applications Conference, 0-78037-420-7USA, October 2002.

20. F. Sahin, A. J. A. Vandenput, 1999Design Considerations of the Flywheel-Mounted Axial-Flux Permanent Magnet Machine for a Hybrid Electric Vehicle, Proceedings of 8th Power Electronics and Applications Conference, 9-07581-506-9Switzerland, September 1999.

21. Semikron 2010 Application Manual", available from http://www.semikron.com/skcompub/en/application_manual-193.htm.

22. S. E. Skaar, R. Nielssen, 2003Genetic Optimization of Electric Machines, a State of the Art Study, Proceedings of 10th Power Electronics and Applications Conference,9-07581-506-9France, September 2003.

23. K. Socha, 2008Ant colony optimization for continuous and mixed-variable domains, PhD Thesis, Université Libre de Bruxelles.

24. K. Tseng, J. , G. H. Chen, 1997Computer-Aided Design and Analysis of Direct-Driven Wheel Motor Drive. IEEE Transactions on Power Electronics, 12 3May 1997), 517 527 0885-8993

25. C. Versèle, et al. 2009Analytical Design of an Axial Flux Permanent Magnet In-Wheel Synchronous Motor for Electric Vehicle, Proceedings of 13th Power Electronics and Application Conference, 978-1-42441-742-1Barcelona, Spain, September 2009.

26. C. Versèle, O. Deblecker, J. Lobry, 2010Multiobjective Optimal Design of Medium Frequency Transformers for Full-Bridge DC-DC Converters, International Review of Electrical Engineering, 5 4July-Augustus 2010), 1354 1363 1827-6660

27. Y. Yang, P. , Y. Luh, P. , C. Cheung, H. , 2004Design and Control of Axial-Flux Brushless DC Wheel Motors for Electric Vehicles- Part I: Multiobjective Optimal Design and Analysis. IEEE Transactions on Magnetics, 40 4July 2004), 1873 1882 0018-9464

28. Y. Yang, P. , D. Chuang, S. , 2007Optimal Design and Control of a Wheel Motor for Electric Passengers Cars. IEEE Transactions on Magnetics, 43 1January 2007), 51 61 0018-9464

29. P. L. Yu, 1973A class of solutions for group decision problems, Management Science,19 8April 1973), 936 946 0025-1909

CHAPTER 3

Design and Analysis of Multi-Node CAN Bus for Diesel Hybrid Electric Vehicle

Xiao Jian Mao[1], Jun hua Song[1], Junxi Wang[1], Hang bo Tang[1] and Zhuo bin[1]

[1] School of Mechanical Engineering, Shanghai Jiaotong University, China

1. INTRODUCTION

Automobile industry will face great evolution in the 21st century. Developing Energy saving and low emission products become the two directions of automobile industry. People have begun to focus on HEV since 1970s. HEV integrates power devices such as engine, motor, battery, which allow its both strong points of pure electric and traditional automobile. CAN is a serial communication protocol initially developed by BOSCH [1]. CAN supports distributed real-time control applications with dependability requirements. CAN is used widely in automotive electronics, with bit rates up to 1 Mbit/s [2-4]. CAN is a multi-master, broadcast protocol with collision detect and a resolution mechanism based on message priorities. Each message on the CAN bus has a unique priority and is only transmitted from a single node on the bus. J1939 is a protocol developed by SAE[5]. The J1939 protocol is a vehicle application layer built on the CAN protocol. The central entity is the Protocol Data Unit (PDU), which carries all the important information needed for determination of a message's priority and size.

In this paper, a multi-node CAN bus of diesel hybrid electric vehicle is designed, based on CAN2.0 protocol and J1939. The design methods of hardware and software for CAN bus are presented.

2. SYSTEM OVERVIEW

In general, according to powertrain configuration, HEV can be classified into three types, namely serial hybrid, parallel hybrid and serial parallel hybrid electric vehicle. In this study, parallel hybrid electric vehicle is adopted, as shown in Fig.1. The motor here is Integrated Starter Generator（ISG）. This design has many advantages such as better inherited of former buses in structure, facility in application and wider applied range. And this method is applied widely in China.

3. MULTI-NODE CAN BUS FOR HEV

3.1. Analysis Of Can Topology

According to requirement of HEV, CAN communication regular among ECU is worked out. CAN bus topology structure is designed. Hierarchical control method is used in the HEV controlled system, and the kernel controller of this powertrain is Hybrid Control Unit (HCU). The controllers of lower lever include Engine Management System (EMS), which is a diesel controller in this system, Battery Packets Control Module (BPCM), Automation Disconnect Module (ADM), Drive Motor Control Module (DMCM). And the main communication between controllers is CAN communication. Main CAN nodes of HEV powertrain are shown Fig. 2. HCU receives information form other lower ECU in order to know the whole vehicle state, at the same time HCU sends control massage to them.

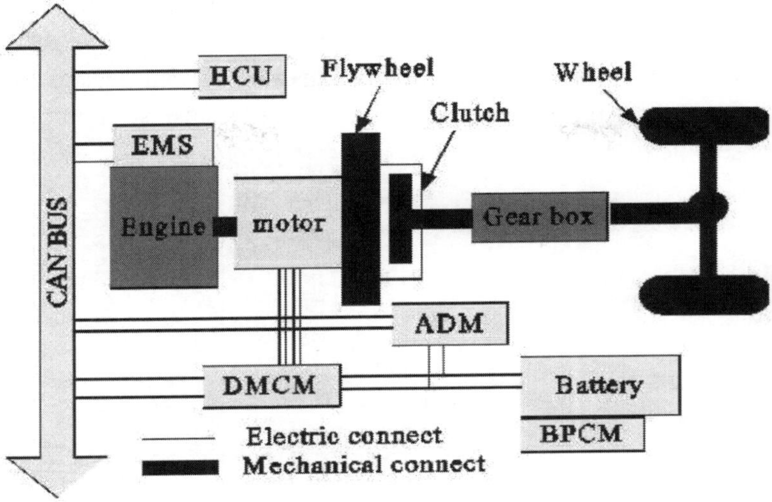

Figure 1. Structure of Diesel Hybrid Electric Vehicle

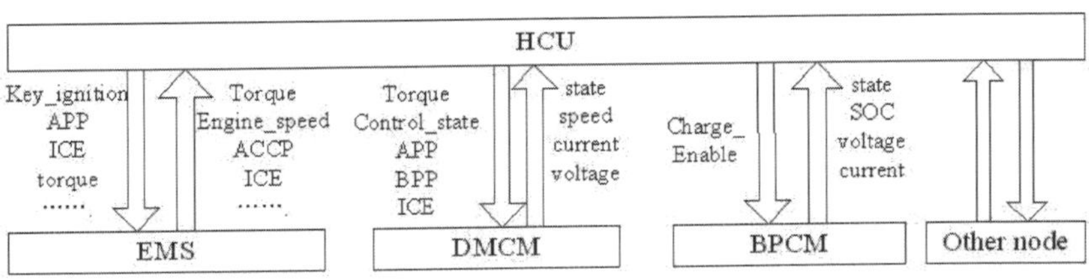

Figure 2. Hierarchical control diagram of hybrid vehicle powertrain system

3.2. Design Of Can Bus for Hybrid Electric Vehicle

HCU is designed based on the microprocessor MC68376[6], and multi-node CAN is developed based on 29bits extended frame. This MCU integrates one TouCAN module which meets CAN2.0B protocol, and this module has 16 buffers. Specific address is assigned for each CAN node, so it is not necessary for declare and modification of each ECU. The address is defined when HCU under power on reset. The information of CAN frame is shown in the table 1. There are two trigger methods for CAN frame, one is period trigger mode, the other is interrupt trigger mode. The period of every node is designed based on the requirement of sampling velocity for vehicle control. Each node has different transmitting and receiving (TR) period. This method can full the control requirement and reduce the road rate of CAN bus, so improve the response time of the system.

Table 1. Information of CAN frames

CANbuffer	Use	Communiaca-tion period
buffer0	HCU→DMCM	20ms
buffer1	HCU→BPCM	1000ms
buffer3	HCU→ADM	20ms
buffer4	DMCM1→HCU	20ms
buffer6	DMCM2→HCU	50ms
buffer7	HCU→Calibration	interrupt trigger
buffer8	BPCM1→HCU	1000ms
buffer9	BPCM2→HCU	1000ms
buffer10	ADM→HCU	20ms
buffer11	Calibration→HCU	Interrupt trigger
buffer12	HCU→DCN1/ LFE→HCU	20ms /100ms
buffer13	HCU→DCN2/ ET→HCU	20ms /100ms
buffer14	HCU→DCN3	20ms
buffer15	EEC3→HCU	50ms

Table 2. Detail data format of DMCM to HCU

Byte position	Data definition
1	DM_State_Flg bit1-4 DM_Dig_Flg bit5-8
2	High byte of DM_Trq_Actual
3	low byte of DM_Trq_Actual
4	high byte of DM_Spd_Actual
5	low byte of DM_Spd_Actual
6	high byte of DM_I_Actual/
7	low byte of DM_I_Actual
8	Set FF

Data combination and data split are applied, when data's addresses were defined. For example, the detail data definition is shown in Table2. Some variables which have two or three bits are combined in one same byte, while some variables which occupy two bytes were split high byte and low byte which in two bytes. These methods can improve the ability of CAN transmitting, which can save the space of transmitting. In addition, the load rate of CAN bus could be reduced.

4. HARDWARE AND SOFTWARE DESIGN OF CAN COMMUNICATION

4.1. Design Of Hardware Circuit Of Can Communication

TouCAN module integrated MC68376 and CAN transceiver 82C250 are adopted in the hardware design of CAN circuit. In order to improve the reliability of communication, power isolation and optoelectronic isolation are applied in the hardware design of CAN communication circuit. DC/DC isolation circuit module is used for power isolation. 6N137 is applied for optoelectronic isolation. The block diagram of CAN circuit is shown in Fig.3. Anti-jamming design should be considered in hardware design. Shield wire, impendence match of bus and electromagnetic filtering are applied in hardware design to improve the communication quality.

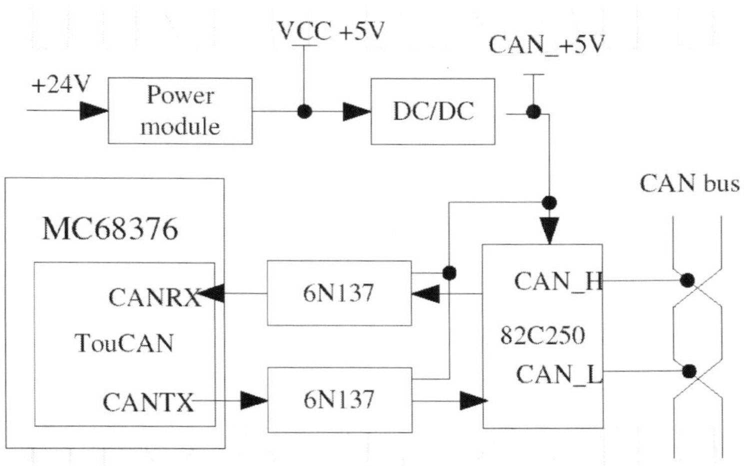

Figure 3. Block chart of CAN circuit

4.2. Design Of Can Communication Software Algorithmic

4.2.1. Buffer Time-Sharing

TouCAN module that meets CAN2.0B protocol, has sixteen buffers. There are eighteen data frames in our control system. In this study, a method of buffer sharing is designed, which two frames use the same buffer, named buffer time sharing. For example, the frames of HCU→DCN1 and LFE→HCU use the same buffer twelve. The flowchart of this the soft structure is shown in the figure5. Buffer time-sharing applies the mechanism of arbitration of CAN bus to avoid data conflict. What's more, buffer time-sharing can approve the hardware usage of HCU.

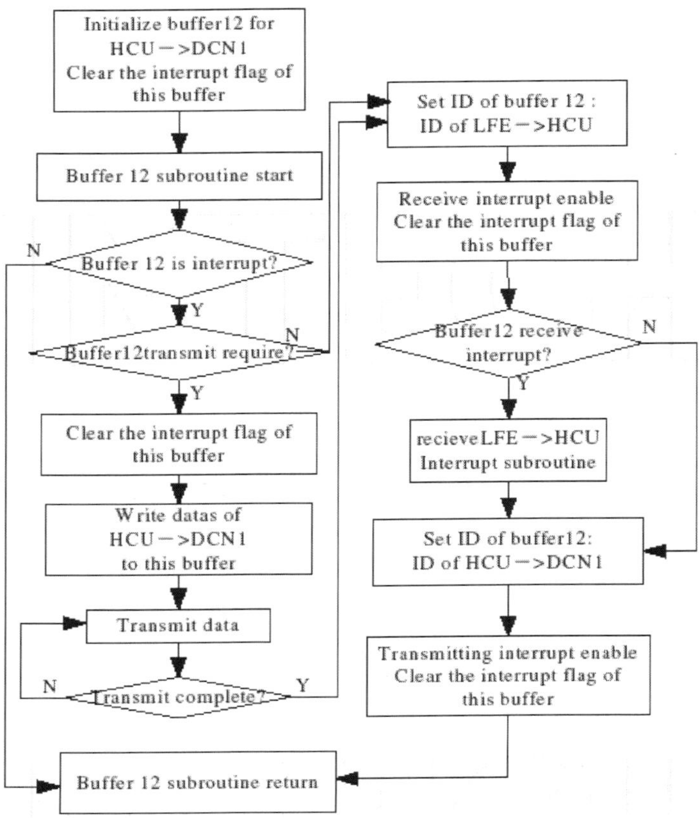

Figure 4. Flowchart of buffer time-sharing program

4.2.2. Time-Sharing Receiving and Transmitting

The infra program of CAN is designed according to HCU. The method of time-sharing receiving and transmitting (TSRT) is applied to ensure quality of communication. The communication period is determined according to requirement of control system. TSRT is an effective way to improve the efficiency of communication. On one hand, TSTR could shorten the running time of CAN subroutines because of all timer subroutine don't achieve at same time. On the other hand, amount of data transmitted on CAN bus is changed, according to time. So the road rate of CAN bus is changed at different time. When there is one timer subroutine operated, the road rate is lower. This is the most ordinary condition.

4.2.3. Optimization Design Of Can Drivers

CAN communication module is one of the most important subroutine of infra program. As an interrup subroutine, CAN driver subroutine may lead to longer time of interrupt relay and affect the ability of interrupt performance of system. A Real Time Operating System (RTOS) is needed in optimization of the control system software. Multi-task is applied in the CAN subroutine design.

The setting of response time of parameters should consider these two conditions.

1. Interrupt trigger is used in the events that need handle immediately. For example, control instruction between HCU and EMS, and calibration instructions between HCU and calibration tools.

2. Considering the response time of control parameters are different, as shown in Table.3. CAN messages are divided into four degrees, 20ms, 50ms, 100ms, 1000ms. At the same time, the priorities of these tasks raise when communication period become short.

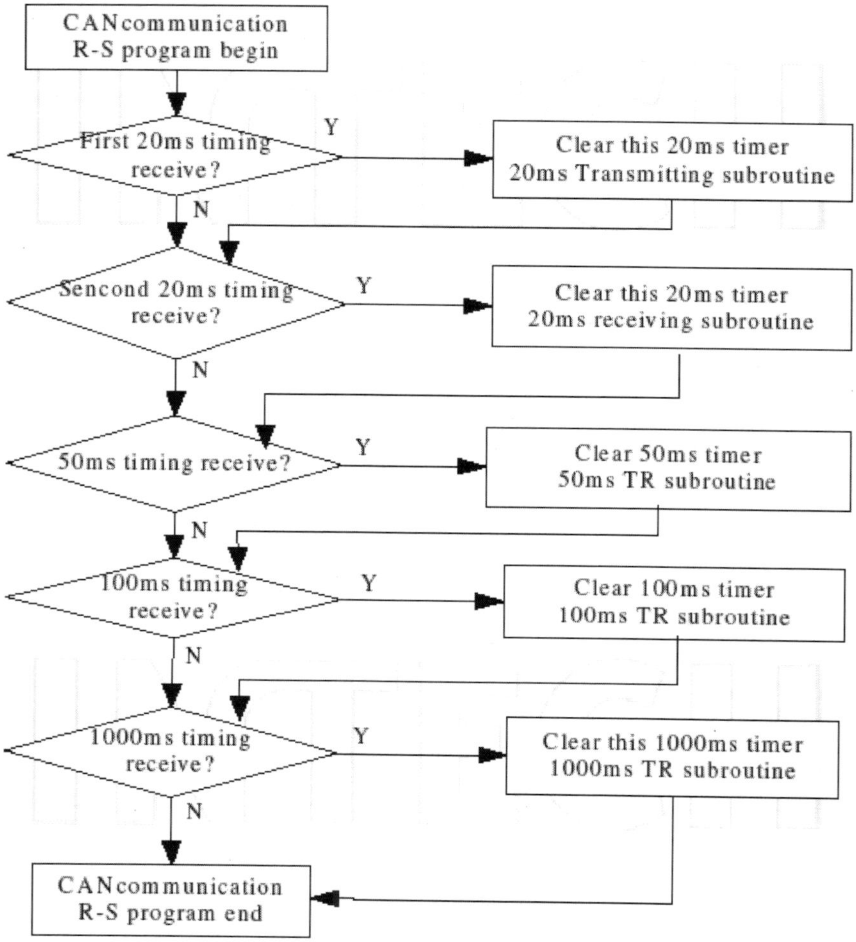

Figure 5. Flowchart of time-sharing program

Table 3. Response time of control parameters

Parameter	response time
Pedal signal	10ms
Brake signal	10ms
Speed of engine	20ms
Speed of vehicle	1000ms
State of charge(SOC)	1s
Torque of driver motor	20ms
Torque of engine	20ms

The new structure of CAN communication is shown in Fig.6. Priority assignment is applied in this new structure. Each task is triggered by event interrupt. If there are two or more tasked triggered at the same time, the task of higher priority is responded firstly. Multi-task method can optimizes the resource of HCU, and supplies an effective means to extend CAN communication program.

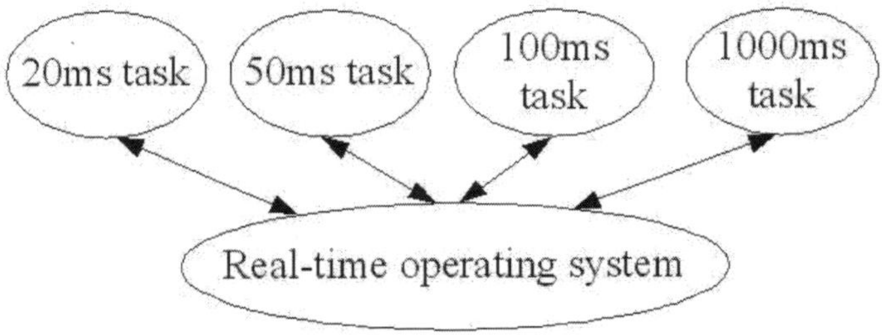

Figure 6. Structure of CAN communication in the RTOS

4.3. Can Bus Road Rate Analysis

The average road rate of CAN bus is defined as the sum total of each communication flame transmitting occupy in bus. It is conducted one equation as following:

$$U_t = \sum_{j=1}^{m} (S_j \times \tau) / T_j \qquad (1)$$

In which, U_t is road rate of whole bus, T is TR period of each communication flame, τ is bit time of CAN data transmitting, Sm is maximum number of TR data bit, m is the number of all communication flames. According to CAN2.0B and J1939, the data can be calculated by Table4.

If there are I bytes data in one CAN frame, form the definition and specification of CAN 2.0B, the maximum number of stuff bits of one frame can be calculated by equation as follow:

$$\frac{54+8i}{5} \quad (2)$$

The maximum bits of each flame TR can be calculated by following equation [7]:

$$s = \frac{54+8i}{5} + 67 + 8i \quad (3)$$

In which, i is the bytes data of each flame TR. The bit time τ is 4us, when baud rate is 250Kb/s. Fromequation (1) and equation (2), the maximum road rate can be calculated is about 28.05%, which is enough for bus communication.

Table 4. Data Frame Of Can Message

	Extend CAN (number of bits)	Bit stuffing?
Start of Frame	1	yes
Arbitration Field	11+1+1+18+1=32	yes
Control field	6	yes
Data field	8*i	yes
CRC	15	yes
CRC delimiter	1	no
ACK field	2	no
EOF	7	no

4.4. Can Calibration Protocol (Ccp)

One of important parts for multi-node CAN bus design is CCP driver. CCP is a CAN based application protocol for calibration and measurement data acquisition of ECUs. It is accepted by many corporations such as VECTOR, DSPACE, ETAS, and become standardization.

CCP driver is needed for ECU which parameters can be calibrated. A CCP driver software is integrated in the infra program of HCU. A basic implementation of CCP driver needs only a few control unit resources such as RAM, ROM, and CPU time. The CCP driver occupies two CAN buffers, which should be assigned low bus priority to avoid influencing other bus communication [8]. The trigger mode of CCP driver is interrupt mode. CCP driver is the foundation of calibration platform. The calibration platform structure is shown as Fig.7. CCP driver achieves data upload, download and calibration of controlled parameters. This calibration system provided reliable, accurate and quickly CAN communication between calibration platform and ECU. It has been used successfully in HEV controlled system.

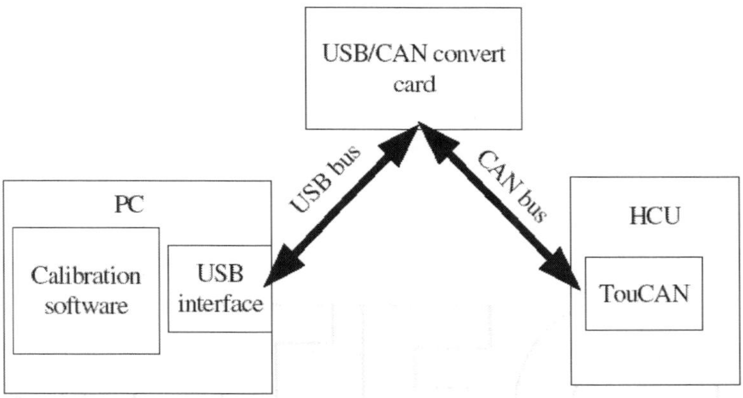

Figure 7. Structure of CCP Calibration platform

5. VERIFY

The verify tests of CAN bus include three parts: Single module verify, hardware in loop (HIL) simulation test, and bench test.

5.1. Single Module Verify

The structure of single module verification is shown as Fig.8.CAN transmitting and receiving (TR) program of single module is carried out in personal computer (PC). The communication between USB/CAN card and single ECU, can verify the CAN hardware function and transmitting and receiving subroutine of the ECU. Single module verification can find the error of hardware or infra TR program as early as possible, which makes well preparation for subsequent tests.

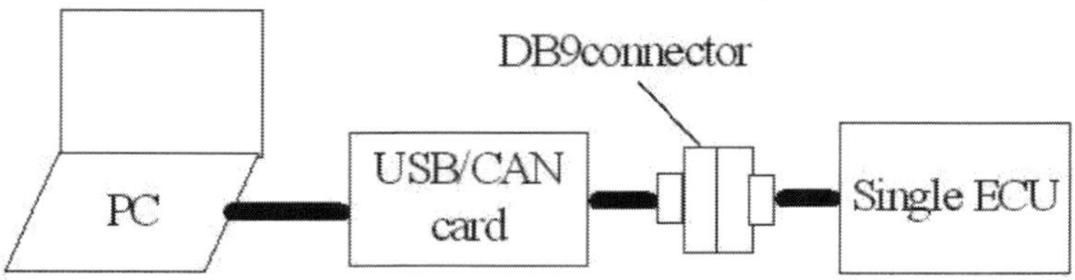

Figure 8. Structure of single control module

5.2. Hardware in The Loop Test

Hardware in the loop test is one of important steps in controller design [9]. The structure of HIL simulation is shown in Fig.9. In HIL system are divided into two important parts: Simulation ECU and upper program of PC. The models of HEV (include engine model, motor model, battery model, ADM model and vehicle model) and monitoring interface are handled in the PC. Communication between HCU and simulation ECU is CAN communication, which uses USB/CAN converter card. The interface of monitoring is developed by Labview. This method of HIL has many advantages. It avoids double RAM communication. The ability of PC could be achieved as vehicle models that are computed in the PC. Multi-thread technology is used in the program design to reduce the response time of communication.

Hardware function can be tested by HIL test. CAN data can display clearly in the monitoring interface. So data communicated can be checked easily. And debug schedule can be quickly. HIL test shows that HCU operates normally, and sampling of sensor's signals are real time, output signals are accurate, and the CAN communication is reliable.

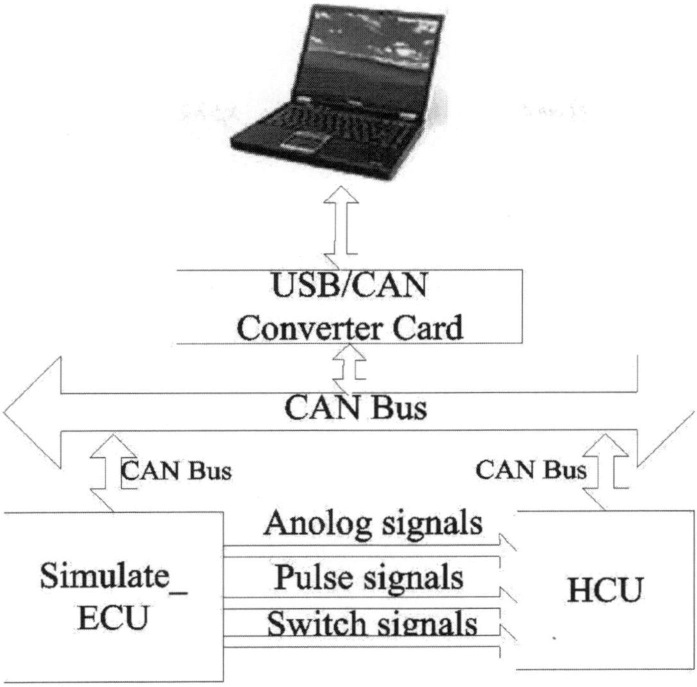

Figure 9. Block diagram of hardware in the loop

5.3. Bench Test

The multi-node test of CAN bus is applied in the bench test of HEV. HCU, EMS, ADM, DMCM and BPCM are concluded. Each ECU simulates the HEV running state, the CAN communication is tested in HEV bench. Figure 10 shows one curve of bench test, which include stages of motor starts engine and motor helps engine.

A reliable CAN bus could be verified from the engine running and monitoring data. The results of bench test shows that CAN hardware circuit has reliability and ability of anti-jamming. At the condition of 2600 rpm engine speed, four CAN nodes work normally. The real time communication between HCU and EMS or BPCM or other node can be achieved. What's more, the calibration instruction and monitoring instruction can be realized regularly. The CAN communication is reliable and real time.

6. CONCLUSION

A multi-node CAN bus is designed based on CAN 2.0B and J1939. The structure of HEV control system is introduced. According to the requirement of control system, a CAN topology is developed. The methods of hardware and software design of CAN bus are discussed. Based on the 32-bit MC68376, power isolation and optoelectronic isolation are applied in the hardware design of CAN communication circuit. The design methods of software algorithmic are emphasized. Buffer sharing and time sharing are applied in the program design of CAN bus. The idea of multi-task is introduced. Task assignment and scheduler could optimize the real time of CAN communication, by event interrupt. Road rate of multi-node CAN bus is analyzed. In order to calibrate the ECU, the CCP driver is designed, which can achieve the data upload and download for calibration.

There are many tests to verify the multi-node CAN bus, single module test, HIL test and bench test. The multi-node CAN bus design can be accomplished quickly through these tests. All tests showed that the hardware and software design of multi-node CAN bus could meet the requirements of HEV system. The quality of CAN communication is reliable and real time.

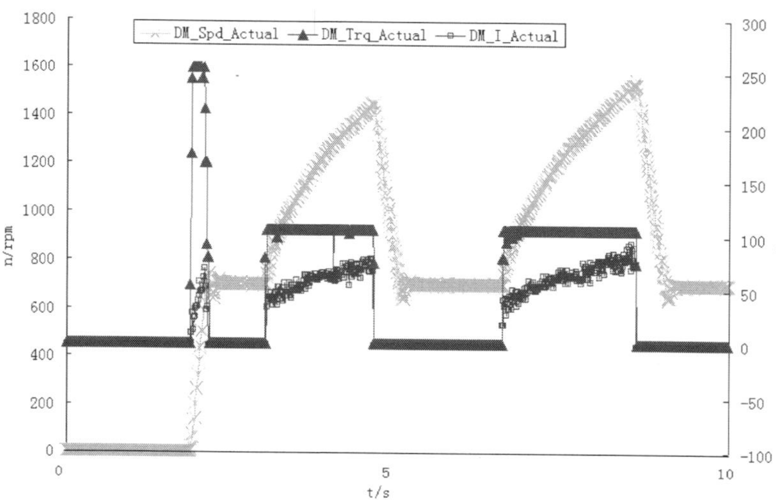

Figure 10. Curve of bench test

7. ACKNOWLEDGEMENTS

The authors could like to acknowledge Xu Jian for his assistance in the project.

REFERENCES

1. B. G. C. A. N. Robert, Version. Specification, 1991
2. C. Massimo, C.. Paolo, 2005CAN_LabView based development platform for fine-tuning hybrid vehicle management systems.IEEE, 433438
3. X. H. Dai, J. X. Zhu, Y. G. Peng, et al.2004of. E. V. Development, system. control, based. communication, C. A. N. on, J]. bus[, Application of Electronic technology, 3739
4. Z. Yao, F. M. Liu, Y. X. Li, 2005Design of Node in CAN Bus for Hybrid Electric Vehicle. Computer and Communications, 121123
5. Society of Automotive Engineering INC.Data Link Layer (1994
6. Motorola INC.MC68336/376 User'sManual.
7. Roger Johansson, Jan Torin.On calculating guaranteed message response times on the SAE J1939 bus. Chalmers Lindholmen University College. Report 10Feb.2002
8. J. -X, J. Wang, X. Feng-J, L. Mao, B. Yang, Zhuo, Development of a new calibration and monitoring system for invehicle electric control units based on controller area network calibration protocol. Proc. IMechE 219PartD: J. Automobile engineering:13811389
9. H. B. Tang, Y. M. Gong, W. C. Tan, B. Zhuo, The hardware design of Hardware-in-Loop Simulation System of Diesel Engine High Pressure Common-Rail ECU Based on CAN Bus. Vehicle Engine.12428

CHAPTER 4

Factors Influencing the Behavioural Intention towards Full Electric Vehicles: An Empirical Study in Macau

Ivan K. W. Lai [1,*], Yide Liu [2,*], Xinbo Sun [3], Hao Zhang [3] and Weiwei Xu [2]

[1] School of Business and Hospitality Management, Caritas Institute of Higher Education, Tseung Kwan O, New Territories, Hong Kong, China

[2] School of Business, Macau University of Science and Technology, Taipa, Macau, China

[3] School of Business Administration, Northeastern University, Heping District, Shenyang 110819, China

ABSTRACT

This study examines the factors that influence individual intentions towards the adoption of full electric vehicles. A sample including 308 respondents was collected on the streets of Macau. The collected data were analysed by confirmatory factor analysis and structural equation modelling. The results demonstrate that environmental concerns and the perception of environmental policy are antecedent factors of the perception of full electric vehicles, which influences the behavioural intention to purchase full electric vehicles. This study also finds that the perception of economic benefit is one of the key factors influencing the adoption of full electric vehicles. Vehicle operators seek economic benefits from future long-term fuel savings, high energy efficiency, and cheap electricity. Thus, a government striving to promote low-carbon transportation needs to scale up its efforts to enhance citizens' environmental concerns and to establish proper environmental policy as well as to provide long-term financial and strategic support for electric vehicles.

KEYWORDS

consumer acceptance; environmental concern; environmental policy; electric vehicles

1. INTRODUCTION

Transportation ranks second after electric power as the largest source of carbon emissions in the world [1]. Over the past few decades, research has been conducted to investigate various aspects of the development of sustainable low-carbon transportation technologies to reduce carbon emissions. As a result, there are already a number of potential alternatives to the conventional diesel/petrol combustion engine [2]. An important development that can improve fuel efficiency and decrease emissions is the introduction of hybrid electric vehicles [3]. However, hybrid electric vehicles are equipped with diesel engines that generate carbon dioxide and cause air pollution. Another alternative is full electric vehicles (energy provided by a battery), which have a zero-emission potential when electricity is produced with the use of renewable energy sources [4]. In fact, powering electric vehicles through solar charging stations could reduce the greenhouse gas (GHG) emissions of these vehicles by up to 34% [5]. Although full electric vehicles have been available since the dawn of motoring, they were not popular. Due to contemporary environmental concerns, full electric vehicles have been making a comeback in the 21st century. Mass-produced full electric vehicles are being introduced into the market by many car manufacturers. For example, the Renault-Nissan alliance sold its 200,000th electric vehicle in early November 2014, approximately four years after the launch of the Nissan LEAF [6]. To support the widespread adoption of full electric vehicles, there is a need to examine the factors influencing the consumer acceptance of these vehicles because consumer acceptance is a key to the commercial success (or failure) of full electric vehicles [7].

There are many factors that influence car-purchasing behaviour, including actual situational factors such as regulatory environments [8]. In addition to the actual situational factors, psychological factors, such as personal attitudes, are equally important [9,10]. Although some empirical studies of the consumer acceptance of hybrid vehicles have been conducted (e.g., [11,12]), there is little research that considers the perception of an expected situation; in particular, there has been little focus on the perception of full electric vehicles.

This study addresses the need for an empirical study that analyses the psychological factors with the situational factors that impact the consumer acceptance of full electric vehicles, and tests the relationships among these factors. This research will identify the factors that influence the consumer acceptance of full electric vehicles and thus might influence policies designed to promote the adoption of full electric vehicles to reduce carbon emissions from transport.

2. LITERATURE REVIEW

2.1. Full Electric Vehicles

Full electric vehicles have been around since before the turn of the twentieth century, and they were popular until approximately 1918 [13]. The continued improvement of the gasoline-powered internal combustion engine vehicles led them to be too competitive [14], and by 1933, full electric vehicles were totally phased out of the transportation market. After a hundred years of evolution, most major vehicle manufacturers are currently developing compact full electric vehicles, usually for short-range city driving (e.g., [15,16,17,18]).

Full electric vehicles have an all-electric drivetrain powered from a battery that is recharged from the electricity supply. The previous generation of full electric vehicles were typically small cars (termed “superminis” in the U.K. and “compacts” in the USA) with a limited range (e.g., 100 km), requiring hours to recharge [19]. Therefore, Cheron and Zins [20] reported that range, speed, recharging time, and dead battery problems were the factors discouraging the purchase of full electric vehicles. In general, the purchasing price of full electric vehicles is higher than that of conventional cars. Past studies concluded that among these disadvantages, limited range is the overwhelming drawback of full electric vehicles [21,22,23,24]. However, this shortcoming has been overcome. Currently, electric vehicles can reach 250 km, have an attractive appearance, and come in a range of sizes. Gerssen-Gondelach and Faaij [4] forecasted that once future Li-ion and ZEBRA batteries could provide sufficient power at a very low cost, full electric vehicles would become competitive with gasoline vehicles. Although the sales of full electric vehicles to date are very low, Weiss *et al.* [25] forecasted that approximately 145 million full electric vehicles will have been produced worldwide by 2035.

2.2. Factors Influencing Alternative Fuel Vehicles Purchasing Behaviour

Twenty years ago, Ellen *et al.* [26] identified the key factors that motivate environmentally conscious behaviours. These factors consist of personal values, such as a concern for the environment and a belief that an individual could make a difference. This study aims to identify the factors that affect the acceptance of full electric vehicles through understanding the factors that influence individual’s purchase intentions for other alternative fuel vehicles. In 2001, the Electric Power Research Institute found that gas prices greatly impact expressions of purchase interest in hybrid electric vehicles [27]. Other customer preferences for hybrid electric vehicles included reduced maintenance, better handling, and reduced air pollution.

Recently, Gallagher and Muehlegger [11] studied the consumer adoption of hybrid electric vehicles in the USA and found that groups with strong preferences for environmentalism and energy security prefer hybrid electric vehicles. Their results indicate that rising gasoline prices and certain social preferences result in maximum sales. Musti and Kockelman [12] found that the top three attributes that buyers look for when seeking a new vehicle purchase are price (30%), fuel economy (28%), and reliability (21%). In general, car use and car ownership are typically associated with instrumental, hedonic, and symbolic attributes [28]; Schuitema *et al.* [29] found that instrumental attributes are largely important for the adoption of electric vehicles and that people's pro-environmental self-identity has a positive effect on the perception of electric vehicles. Zhang *et al.* [30] analysed the factors impeding the development of electric vehicles and found these factors to include deficient electric vehicles subsidy policies, embarrassed electric vehicles market, local protectionism, and unmatched charging infrastructure. Although Steg *et al.*'s [31] survey in the Netherlands suggested that socio-economic and socio-demographic variables are motivational factors for car use, the results of Delang and Cheng's [32] survey in Hong Kong indicated that people recognize the environmental benefits of electric cars but not the economic and social benefits.

For the negative factors, EPRI [33] found that consumers identified the following to be barriers to electric vehicles purchase: lack of electric vehicle infrastructure; potential increases in electric rates; and lack of choice in vehicles. The results of Tan *et al.*'s [34] survey of customer preferences and acceptance of electric vehicles indicated that purchasing behaviours are affected by four factors: charge inconvenience, short battery range, cost, and psychological factors. Bockarjova and Steg [35] claimed the most important barriers for electric vehicle adoption were perceived high monetary and non-monetary costs of electric vehicles and benefits associated with the use of a conventional vehicle. Liu and Santos [36] had similar findings in China that when deciding whether to buy a hybrid electric vehicle, a consumer will consider the cost of the vehicle and the cost of operations such as battery capacity and possible speeds. Other than the price concern, Carley *et al.*[37] argued that range and charge time are barriers to deciding to purchase an electric vehicle in the U.S.A. However, Skippon [38] stated that the future dynamic performance and cruising performance of electric vehicles might partially offset the reduced utility of low range, long recharge times, and higher costs.

However, Klockner *et al.* [39] found that psychological determinants show a high correlation between the purchase and use of electric vehicles in Norway. Nayum *et al.* [40] and Nayum and Klockner [41] further tested the effects of socio-psychological and psychological factors on the normative and intention processes of purchasing an electric vehicle. Klockner [42] further identified the psychological determinants in different stages (goal, behavioural, and implementation stages) of the decision-making process. Their findings confirmed that personal norms, attributes, perceived behavioural control, and planning abilities affect the intention of purchasing an electric vehicle.

Hydrogen fuel cell vehicles (HFCV) are another type of alternative. HFCV and full electric vehicles use electric motors, the only difference is the power source: hydrogen fuel cells *versus* batteries, respectively. Recently, Kang and Park [43] reported that experience with HFCV, policy experience, perception of HFCV, perception of policy, values, and psychological needs are the factors that influence consumer acceptance of HFCV. Tarigan *et al.* [44] and Tarigan and Bayer [45] argued that environmental attitudes and knowledge are important factors for the acceptance of HFCV.

Table 1 summarizes the results of previous studies on the factors influencing purchasing behaviour for alternative fuel vehicles. Some factors are psychological, such as environmental concerns [19,46,47,48] and the negative perception of electric vehicles [19,43], whereas other concerns include actual situational factors, such as tax reduction [49,50] and changes in gasoline prices [11,48,51]. However, little research has been conducted on the effect of the perception of expected situations on the consumer acceptance of alternative fuel vehicles and investigated the interrelationships among them.

Table 1. Factors influencing alternative fuel vehicles purchasing behaviour.

Author(s)	Factor	Vehicles
Cheron and Zins [20]	Expectancies: comfort, reliability, durability, power, road handling, safety, economy, and fair price of parts; Perceived risks: out of power, having an accident, a mechanical breakdown, not being able to start up, being stuck in a traffic jam, and having a flat tire	Full electric vehicles
Chiu and Tzeng [52]	Purchasing price, reliability, maximum speed, emissions level, operating cost, style, agility, safety, cruise distance, and acceleration	Full electric motorcycles
Lipman and Delucchi [53]	Manufacturing costs, retail prices, and lifecycle costs	Hybrid electric vehicles
Sallee [50]	Tax credits	Hybrid vehicles
West [54]	Gasoline prices	Sport utility vehicles
Chandra *et al.* [49]	Tax rebates	Hybrid vehicles
Author(s)	Factor	Vehicles
Diamond [55]	Incentive, income, high occupancy vehicle, gas, VMT (annual cost of fuel), green planning capacity	Hybrid vehicles
Berensteanuand and Li [51]	Gasoline prices, government support	Hybrid vehicles
Gallagher and Muehlegger [11]	Government incentives (tax incentives), changes in gasoline prices, and preferences for environmentalism	Hybrid vehicles
Kang and Park [43]	Experience with HFCV, experience of policy, perception of HFCV, perception of policy, values, and psychological need	Hydrogen fuel cell vehicles (HFCV)
Zhang *et al.* [46]	Understanding of AFVs, vehicle performance, government policy, environmental requirement, opinion of peers, vehicle price, tax reduction, fuel price, fuel availability, maintenance costs, and vehicle safety	Alternative fuel vehicles (AFVs)
Graham-Rowe *et al.* [19]	Cost minimization, vehicle confidence, vehicle adaptation demands, environmental beliefs, impression management, and perception of electric vehicles	Full electric vehicles and hybrid electric vehicles
Tarigan *et al.* [44]	Knowledge, personal profit, and environmental attitude	Hydrogen vehicles
Tarigan and Bayer [45]	Pro-environmental attitudes and hydrogen knowledge	Hydrogen vehicles
Ziegler [47]	Purchase price, motor power, fuel costs, CO_2 emissions, and service station availability	Alternative energy vehicles
Carley *et al.* [38]	Environmental views index, owns a hybrid, appearance, charging stations, range, vehicle price, charge time, car for the environment, innovation, independent on foreign oil	Plug-in electric vehicles
Jensen *et al.* [48]	Purchase price, fuel costs, driving range, carbon emissions, top speed, battery stations, battery life, charging, environmental attitude	Internal combustion *vs.* electric vehicles
Klockner *et al.* [39]	Awareness of need, ascription of responsibility, social norm, descriptive norm, introjected norm, personal norm, perceived behavioural control, awareness of consequences, attitude, intention	Normal *vs.* electric vehicles
Nayum *et al.* [40]	Socio-demographic factors, norm-related items, specific attitude, environmental attitude, intention, perceived behavioural control, brand loyalty, car type class, CO_2 emission levels	Environmental friendly cars
Schuitema *et al.* [29]	Instrumental, hedonic, symbolic, pro-environmental identity, car-authority identity	Normal vehicles, hybrid electric vehicles, and fully electric vehicles
Bockarjova and Steg [35]	Severity (env. & energy), vulnerability (env. & energy), rewards, self-efficacy (env. & energy), resp-efficacy, costs	Electric vehicles
Klockner [42]	Awareness of need, responsibility, personal norms, attitudes, perceived behavioural control, knowledge, planning ability, intentions	Electric vehicles
Nayum and Klockner [41]	Perceived behavioural control variables, importance of car attributes, norm activation constructs, norm-related items, ecological worldview, attitudes, intention, knowledge of environmental impacts, extended socio-demographics	Internal combustion engine *vs.* electric vehicles
Peters and Dutschke [56]	Relative advantages, compatibility, ease of use, trialability, observability, social norm	Electric vehicles
Skippon [38]	Dynamic performance, cruising performance	Electric vehicles

2.3. Full Electric Vehicles in Macau

Macau is a tourism city. Air quality is a pivotal factor in people's travel decisions. Macau does not have power stations. It imports electricity from the China Southern Grid (CSG) supplied by coal generation in Guangdong, so vehicles are the major source of carbon emissions that cause air pollution in Macau. Macau citizens perceive the advantage of a reduction in air pollution and carbon emissions from vehicles. Thus, promotion of alternative fuel vehicles, especially full electric vehicles, has become part of the Macau government's environmental policies [57,58]. This study focuses on the consumer acceptance of full electric vehicles in Macau.

Limited range is one of the weaknesses of full electric vehicles [55]. However, Macau is a small city (total area = 31.3 km^2) with narrow roads and streets (total length = 413.4 km). There were 240,107 vehicles among the population of 636,200 on 31 December 2014 [59], and therefore, one of every 2.65 citizens has a vehicle. Compared with other Asian cities, this ratio is high. The typical driving range of vehicles in Macau is less than 40 km (private usage: 8 km; business usage: 20 km; special usage such as fast-food delivery: 40 km) [60]. Therefore, full electric vehicles are highly suitable for Macau's topography and environment. After visiting Macau, many experts have affirmed the city's suitability for full electric vehicles and have invariably said that Macau is a role model for the implementation of full electric vehicles [61]. The adoption of full electric vehicles could also become an eye-catching way to enhance and re-brand the tourist experience in Macau. Companhia de Electricidade de Macau (CEM) acquired a full electric car and built the first public electric vehicle charging station in early 2010. There are two companies importing full electric vehicles, Mitsubishi and Tesla. BMW plans to announce its full electric vehicle (Mini E) in the coming years. Compared with other tourism cities, Macau appears to have a great potential market for full electric vehicles. The experience of adopting full electric vehicles in Macau could provide guidance for the governments of other cities in developing policies to promote the adoption of full electric vehicles. Although Macau is a small city in Asia, the results of the study could provide insights for comparable cities elsewhere.

3. RESEARCH MODEL AND HYPOTHESES

Some authors have indicated that the effect of tax incentives increases hybrid vehicles' sales (e.g., [11,49,50]). Some authors concluded that the effect of gasoline prices (future fuel savings) affects hybrid vehicles sales (e.g., [11,51,62,63,64]). Recent studies indicated that the perception of government policy and vehicle drivers' environmental attitude affect consumer acceptance of hybrid electric vehicles and HFCV (e.g., [43,45,54]). Most of the previous studies focused only on the acceptance of hybrid electric vehicles and HFCV, but this research investigates the consumer acceptance of full electric vehicles. However, the importance of environmental factors compared with other factors and the relationship between these factors for the adoption of alternative fuel vehicles have not been well studied. Because the market for the full electric vehicle is too new for collecting data to measure actual purchasing behaviour, there may be a gap between intentions and actual behaviour [65]. Thus, this research studies individual intentions towards the adoption of full electric vehicles. In this study, environmental concern and perception of environmental policy are combined with perception of economic benefits, perception of electric vehicles, and behavioural intention towards electric vehicles to form a research model as shown in Figure 1.

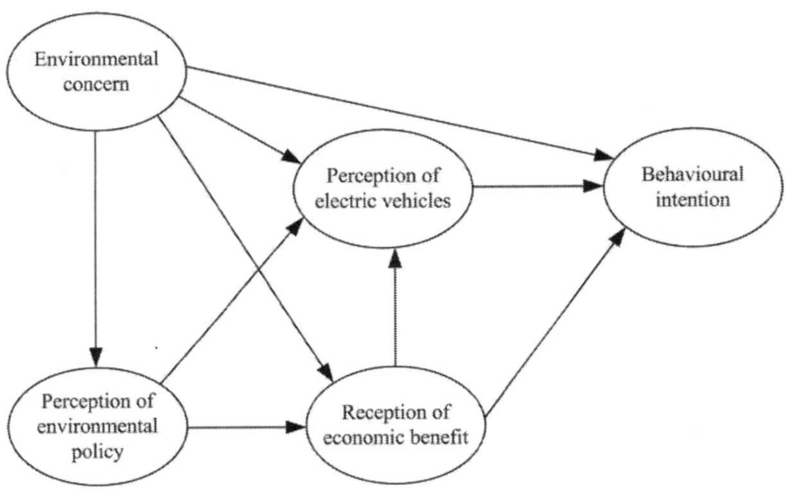

Figure 1. Research model.

Environmental Concern (EC): Environmental concern is usually focused on a perception that the environment is deteriorating in some way [66]. Environmental concern can be defined as the degree to which people are aware of environmental problems and support efforts to solve these problem and/or indicate a willingness to contribute personally to the solution [67,68]. Concern for the environment is significantly related to consumer behaviours such as purchasing ecologically safe products, recycling newspapers, contributing to environmental groups, communication with public officials, and attending public hearings [43]. People with little environmental concern do less work to support environmental policy [69]. Kahn's [70] study indicated that environmentalists are more likely to purchase hybrid electric vehicles than non-environmentalists. Jensen *et al.* [48] argued that environmental concern has a positive effect on the preference for electric vehicles both before and after experiencing an electric vehicle. Peters and Dutschke [56] found that having environmental advantages is a motivator for adopting electric vehicles. Bockarjova and Steg [35] stated that people are more likely to adopt an electric vehicle when they expect electric vehicles to decrease environmental risks. Environmental concern is a psychological factor that should affect user's attitudes towards the acceptance of full electric vehicles.

Hypothesis 1:

A user's environmental concern has a positive impact on the user's perception of environmental policy.

Hypothesis 2:

A user's environmental concern has a positive impact on the user's perception of economic benefit.

Hypothesis 3:

A user's environmental concern has a positive impact on the user's perception of full electric vehicles.

Hypothesis 4:

A user's environmental concern has a positive impact on the user's behavioural intention towards full electric vehicles.

Perception of Environmental Policy (PEP): Irwin and Wynne [71] stated that political context affects the validation of new technology, whereas Flynn and Bellaby [72] argued that political circumstance influences the acceptance of products and technology. There are different types of environmental policies such as financial incentives and sales tax waivers on full electric vehicle purchases. Peters and Dutschke's [56] study indicated that financial incentives are considered to be important measures for purchasing an electric vehicle. Gallagher and Muehlegger [11] and Chandra *et al.* [49] found that tax incentives increase hybrid vehicle adoption. An individual's perception of environmental policy should affect the individual's perception of full electric vehicles.

Hypothesis 5:
A user's perception of environmental policy has a positive impact on the user's perception of economic benefit.

Hypothesis 6:
A user's perception of environmental policy has a positive impact on the user's perception of full electric vehicles.

Perception of Economic Benefit (PEB): The acceptance of a product is often affected by a personal perception of economic benefit. Berensteanu and Li's [51] study indicated that the effect of high gasoline prices increases hybrid vehicles sales compared with gasoline vehicles because the running costs and maintenance costs of full electric vehicles are low. The results of Wang and Gonzalez's [73] study indicated that the energy costs of electric vehicles are approximately eight times less than those of gasoline, diesel, and natural gas vehicles. Expensive gasoline and cheap electricity are great incentives to buy an electric vehicle rather than a gasoline car [74]. Consumers may consider these benefits when they are making a decision regarding purchasing a new vehicle.

Hypothesis 7:
A user's perception of economic benefit has a positive impact on the user's perception of full electric vehicles.

Hypothesis 8:
A user's perception of economic benefit has a positive impact on the user's behavioural intention towards full electric vehicles.

Perception of Full Electric Vehicles (PFEV): A positive perception of a product can make a customer more likely to purchase the product [75]. Schulte *et al.* [2] supported the view that the perception of a product affects customer acceptance. In this study, "the perception of full electronic vehicles" is the perceived overall driving performance, including the comfortable driving, of electric vehicles. Thus, the perception of full electric vehicles should influence customer's purchasing behaviour.

Hypothesis 9:
A user's perception of full electric vehicles has a positive impact on a user's behavioural intention towards full electric vehicles.

4. RESEARCH METHOD

The research question of this study is as follows: what are the factors that affect consumer acceptance of full electric vehicles in Macau? A questionnaire survey was employed that includes two sections. Section 1 contains 5 sets of questions with a total of 15 items for the five constructs of the research model. A 7-point Likert-type scale was employed, with 1 being "strongly disagree" and 7 being "strongly agree". Section 2 provides general background information.

Environmental concern has been treated as an evaluation of or attitude towards one's own behaviour, or other's behaviour with consequences for the environment [76,77,78,79]. The measured items of environmental concern cover air pollution [47,52], environmental problems [11,44], and energy conservation [80]. The potential government policies include sales tax waivers [11,49], the subsidization of the construction of charging stations [47], and full electric vehicle programmes [81]. Long-term economic benefits for utilizing full electric vehicles consist of the fuel savings [47], high energy efficiency [47], and cheap electricity [74]. The measured items employed in the construction of the perception of full electric vehicles are based on Kang and Park's [43] study with some modifications to suit the Macau setting. The behavioural intention in this study is measured as the stated likelihood to purchase a full electric vehicle and to recommend full electric vehicles to friends and others in the future. Thus, the measured items are taken from Zeithaml *et al.* [82]. The measured items are listed in Table 3.

The content validity of the measure was checked by two academic colleagues to assess any misunderstandings or ambiguities of expressions in the questionnaire. To evaluate the readability of the questionnaire, a pilot study with 20 students who owned cars was performed. The feedback from the respondents was that the questions in the questionnaire were easily understood and answerable. In the study itself, the interviewer-administered survey was conducted on the streets at five business and residential areas in Macau in March 2012. A filter question "do you drive your own car?" was asked to qualify respondents. A total of 310 completed questionnaires were collected within a month. However, two questionnaires were eliminated (e.g., for giving the same rating for all items), leaving 308 questionnaires as valid for analysis. The minimum, recommended, and ideal N:q ratios (observations/parameters to be estimated) are 5:1 [83,84], 10:1 [83,84,85], and 20:1 [85], respectively. In this study, there are 39 estimated parameters in the model (as shown in Figure 2), so the N:q ratio is 7.9. Therefore, the sample size of the study is still adequate. Of the sample, 57.8% were male, and 60.7% had a bachelor's or higher education. The respondents in the brackets of 18–29, 30–39, 40–49, and 50 or over accounted for 52.6%, 23.7%, 14.9%, and 8.7% of the total respondents, respectively. The distribution of the samples is similar to that of the population of vehicle owners in Macau because Macau's younger people who work in the tourism sector have a good income and like to drive their own vehicles. Table 2 presents the background characteristics of the respondents.

Table 2. Background characteristics of the respondents (N = 308).

		Frequency	Percent	Cumulative Percent
Gender	Male	178	57.8	57.8
	Female	130	42.2	100.0
Age	18–29	162	52.6	52.6
	30–39	73	23.7	76.3
	40–49	46	14.9	91.2
	50–59	22	7.1	98.4
	≥ 60	5	1.6	100.0
Education	Primary	22	7.1	7.1
	Secondary	99	32.1	39.3
	Undergraduate	169	54.9	94.2
	Postgraduate	18	5.8	100.0
Monthly Income	Less than USD 1875	112	36.4	36.4
	USD1875–3750	171	55.5	91.9
	Over USD 3750	25	8.1	100.0

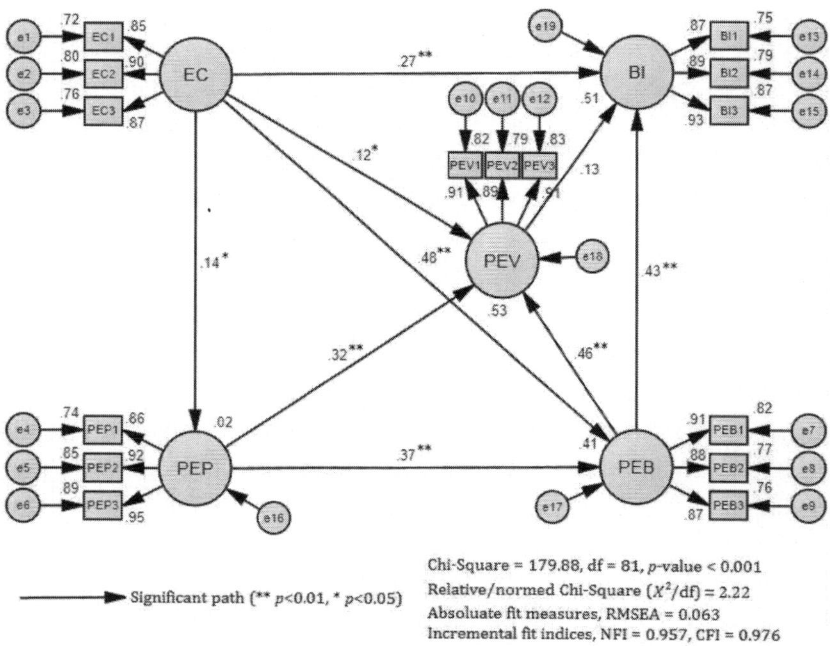

Figure 2. Structural equation modelling results.

5. FINDINGS

Confirmatory factor analysis (CFA) was performed to evaluate the questionnaire's convergent and discriminant validity.Table 3 presents the means, standard deviation, the reliability of the constructs, and the model's standardized factor loadings. The Cronbach's alpha for all components is higher than 0.6, so the reliability of the study is acceptable [86]. All factor loadings are above 0.7 and are highly significant as required for convergent validity ($p < 0.05$). The values of Average Variance Extracted (AVE) of all constructs exceed 0.5, and the values of Composite Reliability (CR) of all constructs exceed 0.7. Based on the guidelines of Hair *et al.* [87], the construct validity of the model is provided. Table 4 indicates the correlation matrix of the five constructs. All of the correlation values among the constructs of the model are significant ($p < 0.01$). The square root of the construct's AVE exceeds its correlations with other constructs in the model, demonstrating a necessary aspect of the discriminant validity of the latent constructs.

The structural equation modelling (SEM) analysis was performed to test the research hypotheses empirically [88,89].Figure 2 presents the results of the SEM analysis. These results suggest that the model provides an acceptable fit [87,90,91] and that all hypotheses are valid. Table 5 presents the direct, indirect, and total effects of the model.

Table 3. Reliability and validity of the constructs.

		Mean	Std. Dev.	Cronbach's Alpha	Factor Loadings	AVE	CR
EC	Environmental concern:	5.553		0.905		0.763	0.906
EC1	I worry about air pollution.	5.507	0.776		0.854		
EC2	I am concerned about environmental problems.	5.490	0.793		0.893		
EC3	I care about energy conservation.	5.662	0.852		0.873		
PEP	Perception of government policy:	4.321		0.933		0.830	0.936
PEP1	I think that the Macau government would waive sales tax for full electric vehicles.	4.497	1.093		0.864		
PEP2	I think that the Macau government would subsidize the construction of charging stations.	4.234	1.045		0.923		
PEP3	I think that the Macau government would announce a full electric vehicles programme to subsidize the cost of electric vehicles.	4.231	0.966		0.944		
PEB	Perception of economic benefit:	5.128		0.916		0.787	0.917
PEB1	I think that electric vehicles could provide the benefit of fuel savings.	5.166	0.681		0.910		
PEB2	I think that electric vehicles could provide the benefit of high energy efficiency.	5.198	0.678		0.875		
PEB3	I think that electric vehicles could provide the benefit of cheap electricity.	5.020	0.744		0.875		
PFEV	Perception of electric vehicles: Adapted from Kang and Park [43].	4.398		0.929		0.809	0.927
PEV1	I think that the riding comfort of electric vehicles would be good.	4.419	0.897		0.903		
PEV2	I think that the driving performance of electric vehicles would be good.	4.351	0.892		0.886		
PEV3	I think that having an electric vehicle would be good.	4.425	0.967		0.908		
BI	Behavioural intention: Adapted from Zeithaml *et al.* [82].	4.932		0.924		0.786	0.917
BI1	I would speak favourably about full electric vehicles to others.	4.990	0.678		0.853		
BI2	I would recommend my friends to buy a full electric vehicle.	4.873	0.651		0.879		
BI3	I would buy a full electric vehicle in the future (say, in 3 years).	4.932	0.634		0.926		

Remark: AVE—Average Variance Extracted, CR—Composite Reliability.

Table 4. Construct correlation matrix.

	Square Root of AVE	EC	PEP	PEB	PEV
EC	0.873				
PEP	0.911	0.155			
PEB	0.887	0.068	0.439		
PEV	0.899	0.230	0.554	0.621	
BI	0.887	0.371	0.328	0.572	0.496

Table 5. Direct, indirect, and total effects.

	PEP	PEB			PEV			BI		
	Direct	**Direct**	**Indirect**	**Total**	**Direct**	**Indirect**	**Total**	**Direct**	**Indirect**	**Total**
EC	0.140	0.481	0.051	0.532	0.123	0.290	0.413	0.270	0.287	0.557
PEP		0.365		0.365	0.320	0.168	0.448		0.225	0.225
PEB					0.461		0.461	0.435	0.062	0.497
PEV								0.135		0.135

Table 6 presents the comparisons of the measurement model, structural model, and three alternative models. This study assumes that environmental concern is an initial antecedent factor and behavioural intention is the final consequence outcome. The first alternative model contains a new path (PEP → BI) (Model A as shown in Figure 3a). The results reveal that PEP does not significantly affect BI. For validating the structure of the model, two alternative models (Models B and C, as shown in Figure 3b,c) were quasi-randomly generated as recommended by MacCallum *et al.* [92]. The results of the SEM analysis indicate that there is a non-significant path in these two models. The model-fit indices of the structural model are better than the model-fit indices of models A and B. Although the model-fit indices of the structural model is the same as the model-fit indices of the model C, in case the non-significant path is removed to form a model D (as shown in Figure 3d), the model-fit indices of model D are worse than the model-fit indices of the structural model. Therefore, compared with the four alternative models, the structural model is the best model, providing the best explanations of the data.

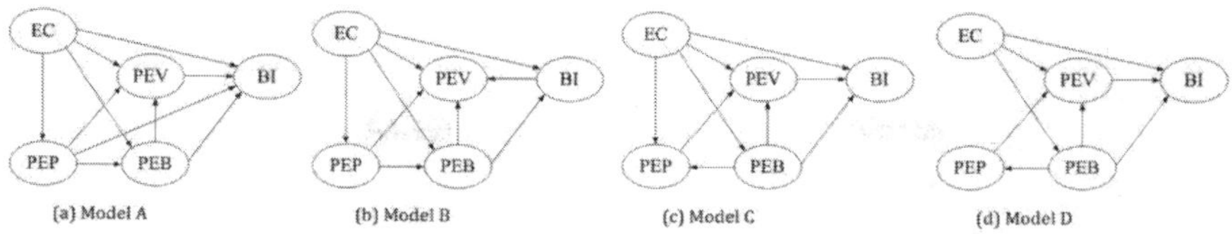

Figure 3. Alternative models.

Table 6. The comparisons of the models.

	Measurement Model	Structural Model	Model A	Model B	Model C	Model D
Chi-Square	179.816	179.879	179.816	180.173	179.879	183.187
Df	80	81	80	81	81	82
χ^2/df	2.248	2.221	2.248	2.224	2.221	2.234
RMSEA	0.064	0.063	0.064	0.063	0.063	0.063
CFI	0.976	0.976	0.976	0.976	0.976	0.975
EC → PEP		0.140 *	0.141 *	0.142 *	−0.126 (ns)	--
PEB → PEP		--	--	--	0.500 ***	0.428 ***
EC → PEB		0.365 ***	0.366 ***	0.480 ***	0.532 ***	0.596 ***
PEP → PEB		0.481 ***	0.481 ***	0.367 ***	--	--
EC → PEV		0.123 *	0.123 *	0.085 (ns)	0.123 *	0.123 *
PEP → PEV		0.320 ***	0.320 ***	0.315 ***	0.320 ***	0.320 ***
PEB → PEV		0.461 ***	0.461 ***	0.394 ***	0.461 ***	0.459 ***
BI → PEV		--	--	0.135 *	--	--
EC → BI		0.270 ***	0.267 ***	0.279 ***	0.270 ***	0.272 ***
PEB → BI		0.435 ***	0.438 ***	0.521 ***	0.435 ***	0.434 ***
PEV → BI		0.135 *	0.142 *	--	0.135 *	0.134 *
PEP → BI		--	−0.014 (ns)	--	--	--

*** $p < 0.001$, * $p < 0.05$.

6. CONCLUSIONS AND DISCUSSIONS

6.1. Theoretical Contribution

This paper takes Macau as a case study to investigate the factors that influence individual behavioural intention towards the full electric vehicle. The results of this study prove that environmental concern is an initial factor that finally leads customer's behaviour intention towards purchase of a full electric vehicle. Environmental concern is a psychological factor that directly and indirectly influences four types of perceptions that mediate the link between environmental concern and the acceptance of full electric vehicles (as shown in Table 5).

Unsurprisingly, the results of this study indicate that the perception of environmental policy is positively correlated with the perception of economic benefit linked to full electric vehicles. The perception of economic benefit and the perception of electric vehicles directly affect the acceptance of full electric vehicles. These findings are consistent with the findings of Kang and Park's [43] study that the perception of government policy directly affects the perception of hydrogen fuel cell vehicles and indirectly affects the acceptance of HFCV. The public is aware of the environmental policy and believes that the policy will be maintained continuously to build a low carbon society.

The results of the data analysis indicate that customers' behavioural intention towards full electric vehicles is affected by environmental concern, the perception of full electric vehicles, and the perception of economic benefit. The SEM analysis shows that the direct effect of the perception of economic benefit ($\beta = 0.435$) is larger than the direct effects of environmental concerns ($\beta = 0.270$) and of the perception of full electric vehicles ($\beta = 0.135$). Many studies have argued that environmental concerns are an initial factor that stimulates the need of environmental light vehicles and environmental policy that encourages consumers to take action towards purchasing environmental light vehicles; however, this study highlights a new insight; the perception of economic benefit is an important factor that affects consumer behaviour towards purchase of full electric vehicles. Graham-Rowe *et al.* [19] stated that vehicle operators are more sensitive to fuel economy, and they attempted to calculate likely costs and savings in fuel consumed. The low cost per kilometre is regarded to be the biggest advantage of full electric vehicles [93]. Although the limited driving range is an obstacle to the adoption of full electric vehicles [94], the results of Daziano and Chiew's [65] study indicated that when vehicle operators expected operating cost savings, they would be satisfied with a short driving range (114.8 miles). Therefore, the perception of economic benefit is an important determinant of the consumer acceptance of full electric vehicles in Macau. However, as most of the studies in technology acceptance, this study measured only behaviour intention rather than real behaviour because, at this moment, full electric vehicles are still not commonly adopted in many countries. Further studies should be conducted to obtain additional data to understand how the perception of economic benefits affects the real purchasing behaviour once full electric vehicles are mass produced.

Many previous studies evaluated actual situational factors for adopting environmental friendly vehicles. If some situations will be realized soon or we want to test the feedback from consumers in some situations (e.g., electric vehicles could benefit from cheap electricity), employing the expected situational factors could help researchers obtain the outcomes of such situations. It can provide a more realistic feedback for policy makers and commercial marketers to consider before taking actions. This research model tests the situational factors (perception of environmental policy and perception of economic benefit) that consumers are looking for. The results of the data analysis indicate that two expected situational factors play important roles in the adoption of full electric vehicles as discussed above. This study also explores the interrelationships among four factors and the user's behavioural intention towards full electric vehicles (as shown in Figure 2). This study contributes a research model that can be further investigated to explain the causal effects of these four factors for the adoption of other environmental technologies.

6.2. Practical Implications

Air pollution harms the tourism economy and is often raised as a concern in the feedback provided by departing guests. To create a green environment, the general public should adopt full electric vehicles in Macau. This study indicates that environmental concern is an antecedent factor that stimulates interest in full electric vehicles. The Macau government should consider educating the public about the importance of environmental protection and the environmental advantages of driving full electric vehicles. In addition to promoting environmental protection concepts through the mass media, the Macau government could provide financial support for environmental organizations to conduct some environmental protection workshops in primary and secondary schools to educate future vehicles' drivers regarding knowledge of full electric vehicles. The Macau government could also consider sponsoring universities in Macau to develop environmental research and monitoring programmes and let college students practice driving full electric vehicles on their campuses.

The introduction of full electric vehicles requires decisive government environmental policies as the Macau government could play a leading role in changing to full electric vehicles. Although the Macau government offers tax incentives for purchasing light environmentally friendly vehicles, these tax incentives also cover hybrid electric vehicles. Like many Western countries, the Macau government's subsidies are not currently aligned with the goal of decreasing gasoline consumption in a consistent and efficient manner [95]. Thus, the Macau government should do more to promote full electric vehicles. The lack of supporting infrastructure may hinder consumer acceptance of full electric vehicles [55]. Many drivers would be more willing to invest in electric vehicles if there were sufficient charging stations in the community [96]; thus, adding recharging locations could increase the proportion of driving done by electric vehicles [97]. The development of charging infrastructure is essential in supporting the broad-based deployment of electric vehicles [98]. Norway is a successful case of promoting electric vehicles where tax exemption, availability of free public charging stations as well as toll-free roads, ferries and the ability of electric-car drivers to utilize bus lanes are important factors that encourage Norwegian drivers to choose electric vehicles [53]. The Macau government should establish the supporting infrastructure for full electric vehicles, such as providing solar powered charging facilities in public car parks, offering free parking for full electric vehicles charged in public car parks, and offering a full electric vehicles programme to subsidize vehicle owners to replace their existing gasoline vehicles with new full electric vehicles.

This study indicates that the acceptance of full electric vehicles as common transportation equipment would be primarily determined by the perception of economic benefits, that is, the long-term cost advantage of full electric vehicles compared with vehicles that utilize gasoline. Consumers care about long-term lifecycle costs [99]. The Macau government should provide long-term financial support. Battery cost is the most important factor for owning a full electric vehicle [4]. Rechargeable batteries are costly because vehicle owners need to spend money replacing drained batteries every few years.

Therefore, the replacement of drained battery would lead to a much higher total cost of ownership [100]. One way around this would be for the Macau government to subsidize full electric vehicle owners for the replacement of drained batteries. However, vehicle maintenance cost is also a long-run cost. Full electric vehicle importers should guarantee that the maintenance costs for full electric vehicles should be competitive with the maintenance costs for vehicles that run on gasoline.

Because Macau does not have a power station, vehicles are one of the major sources of carbon emissions in Macau. Carbon emissions are typically connected to air pollution that poses a threat to the city's main industry—tourism. As a tourist city, the major government income in Macau is from tourism industries. The total income and tax revenues from gaming were above USD17.6 billion and approximately USD 13.4 billion, respectively, in 2013 [59]. The fiscal balance was approximately USD12.4 billion. Since Macau imports electricity from the CSG in Guangdong, so whether both the emissions of greenhouse gases and health-endangering gases could be reduced depends on how the electricity is produced by CSG. The Macau government has relatively large financial reserves and responsibility to support CSG to develop clean energy production, such as building wind energy farms in Hengqin, Zhuhai.

6.3. Limitations and Further Studies

This study focuses only on the citizens and their acceptance of full electric vehicles in Macau. Macau is a small city that is suitable for the adoption of full electric vehicles, and thus, the findings of this study may not be generalizable to other cities with different geographic and economic features. However, the market in Macau is not developed enough to include actual purchase decisions in the model; therefore, further studies are recommended to revise this model for studying the actual purchasing behaviour in a more developed market, for example, in Norway.

As this research investigated only four positive psychological factors, it is suggested that other positive and negative factors such as perceived drawbacks could be added in future studies to widen the scope of this research area. Additionally, there are different types of benefits. This study considers only long-term, personal, direct, and measurable economic benefits that are printed in the catalogues of electric vehicles. However, researchers can consider integrating other benefits such as social and environment benefits in their research.

This study is concerned only with consumer acceptance of full electric vehicles. It may not be generalizable for other environmental transportation technologies. Future studies are suggested to investigate whether a similar concept can be employed with regard to installing other environmental transportation equipment.

ACKNOWLEDGMENTS

The authors gratefully acknowledge the financial support of 2015 Research on Human Resources in Shenyang (A study in breaking the constraints on entrepreneurship and obstacles to innovation), the Macau Foundation, the Faculty Research Grant of Macau University of Science and Technology, and Macao Polytechnic Institute (RP/OTHER-01/2014).

AUTHOR CONTRIBUTIONS

Ivan K. W. Lai developed the research ideas, conducted the research, analysed the data, and wrote the manuscript. Yide Liu implemented the research programme, was involved in some parts of the data analysis, and helped complete the drafting of this article. Xinbao Sun provided guidance and advice during the duration of the project. Hao Zhang and Weiwei Xu distributed the survey and assisted with the process for the project. All authors read and approved the final manuscript.

CONFLICTS OF INTEREST

The authors declare no conflict of interest.

REFERENCES

1 IEA. World Energy Outlook 2006; International Energy Agency: Paris, France, 2006.

2 Schulte, I.; Hart, D.; van der Vorst, R. Issues affecting the acceptance of hydrogen fuel. Int. J. Hydrog. Energy 2004, 29, 677–685.

3 IEA. Technology Roadmap, Electric and Plug-in Hybrid. Electric Vehicles (EV/PHEV); International Energy Agency: Paris, France, 2009.

4 Gerssen-Gondelach, S.J.; Faaij, A.P.C. Performance of batteries for electric vehicles on short and longer term. J. Power Sources 2012, 212, 111–129.

5 Onat, N.C.; Kucukvar, M.; Tatari, O. Towards life cycle sustainability assessment of alternative passenger vehicles. Sustainability 2014, 6, 9305–9342.

6 International Press Website. Renault-Nissan Alliance sells its 200,000th electric vehicle, 26 November 2014. Available online: http://media.renault.com/global/en-gb/alliance/Media/PressRelease.aspx?mediaid=63469 (accessed on 1 July 2015).

7 Egbue, O.; Long, S. Barriers to widespread adoption of electric vehicles: An analysis of consumer attitudes and perceptions. Energy Policy 2012, 48, 717–729.

8 Collins, C.; Chambers, S.M. Psychological and situational influences on commuter-transport-mode choice. Environ. Behav. 2005, 37, 640–661.

9 Choo, S.; Mokhtarian, P.L. What type of vehicle do people drive? The role of attitude and lifestyle in influencing vehicle type choice. Transp. Res. A 2004, 38, 201–222.

10 Laidley, T.M. The influence of social class and cultural variables on environmental behaviors: Municipal-level evidence from Massachusetts. Environ. Behav. 2013, 45, 170–197.

11 Gallagher, K.S.; Muehlegger, E. Giving green to get green? Incentives and consumer adoption of hybrid vehicle technology. J. Environ. Econ. Manag. 2011, 61, 1–15.

12 Musti, S.; Kockelman, K.M. Evolution of the household vehicle fleet: Anticipating fleet composition, PHEV adoption and GHG emissions in Austin, Texas. Transp. Res. A 2011, 45, 707–720.

13 Chan, C.C. An overview of electric vehicle technology. IEEE Proc. 1993, 81, 1202–1212.

14 Bates, B. Electric Vehicles—A Decade of Transition; SAE/PT-40; Society of Automotive Engineers Inc.: Warrendale, PA, USA, 1992.

15 Mitsubishi. Mitsubishi i-MiEV. Mitsubishi Motors Deutschl and GmbH. Available online: http://www.imiev.de/impressum.html (accessed on 30 January 2015).

16 Chambers, N. Nissan Clarifies: 10,000 LEAFs Built by End of March, Faster Charging Upgrade Being "Studied". Plug in Cars, Available online: http://www.plugincars.com/nissan-clarifies-10000-leafs-built-end-march-faster-charging-upgra de-being-studied-106922. html (accessed on 30 January 2015).

17 Citroen. Citroen C-zero. Citroen Deutschland GmbH. Available online: http://www.Citroen.de/Citroen-c-zero/#/Citron-c-zero/ (accessed on 30 January 2015).

18 Nissan. Nissan Leaf. Nissan Nederland. Available online: http://www.nissan.nl/#vehicles/electricvehicles/leaf(accessed on 30 January 2015).

19 Graham-Rowe, E.; Gardner, B.; Abraham, C.; Skippon, S.; Dittmar, H.; Hutchins, R.; Stannard, J. Mainstream consumers driving plug-in battery-electric and plug-in hybrid electric cars: A qualitative analysis of responses and evaluations. Transp. Res. A 2012, 46, 140–153.

20 Cheron, E.; Zins, M. Electric vehicle purchasing intentions: The concern over battery charge duration. Transp. Res. A1997, 30, 235–243.

21 Beggs, S.D.; Cardell, N.S. Choice of smallest car by multi-vehicle households and the demand for electric vehicles.Transp. Res. A 1980, 14, 389–404.

22 Hensher, D.A. Functional measurement, individual preference and discrete-choice modelling: Theory and application.J. Econ. Psychol. 1982, 2, 323–335.

23 Calfee, J.E. Estimating the demand for electric automobiles using fully desegregated probabilistic choice analysis.Transp. Res. B 1985, 19, 287–301.

24 Bunch, D.S.; Bradley, M.; Golob, T.F.; Kitamura, R.; Occhiuzzo, G.P. Demand for clean-fuel vehicles in California: A discrete-choice stated preference pilot project. Transp. Res. A 1993, 27, 237–253.

25 Weiss, M.; Patel, M.K.; Junginger, M.; Perujo, A.; Bonnel, P.; van Grootveld, G. On the electrification of road transport—Learning rates and price forecasts for hybrid-electric and battery-electric vehicles. Energy Policy 2012, 48, 374–393.

26 Ellen, P.S.; Wiener, J.L.; Cobb-Walgren, C. The role of perceived consumer effectiveness in motivating environmentally conscious behaviors. J. Public Policy Mark. 1991, 10, 102–117.

27 EPRI. Comparing the Benefits and Impacts of Hybrid. Electric Vehicle Options; Technical Report 1000349; Electric Power Research Institute: Palo Alto, CA, USA, 2001.

28 Steg, L. Car use: Lust and must. Instrumental, symbolic and affective motives for car use. Transp. Res. A 2005, 39, 147–162.

29 Schuitema, G.; Anable, J.; Skippon, S.; Kinnear, N. The role of instrumental, hedonic and symbolic attributes in the intention to adopt electric vehicles. Transp. Res. A 2013, 48, 39–49.

30 Zhang, X.; Rao, R.; Xie, J.; Liang, Y. The current dilemma and future path of China's electric vehicles. Sustainability 2014, 6, 1567–1593.

31 Steg, L.; Geurs, K.; Ras, M. The effects of motivational factors on car use: A multidisciplinary modelling approach.Transp. Res. A 2001, 35, 789–806.

32 Delang, C.O.; Cheng, W.T. Hong Kong people's attitudes towards electric cars. Int. J. Electr. Hybrid Veh. 2013, 5, 15–27.

33 EPRI. Characterizing Consumers' Interest in and Infrastructure Expectations for Electric Vehicles: Research Design and Survey Results; Report No. 1021285; Electric Power Research Institute: Palo Alto, CA, USA, 2010.

34 Tan, Q.; Wang, M.; Deng, U.; Yang, H.; Rao, R.; Zhang, X. The cultivation of electric vehicles market in China: Dilemma and solution. Sustainability 2014, 6, 5493–5511.

35 Bockarjova, M.; Steg, L. Can Protection Motivation Theory predict pro-environmental behavior? Explaining the adoption of electric vehicles in the Netherlands. Glob. Environ. Chang. 2014, 28, 276–288.

36 Liu, J.; Santos, G. The plug-in hybrid electric vehicles potential for urban transport in China: The role of energy sources and utility factors. Int. J. Sustain. Transp. 2015, 9, 145–157.

37 Carley, S.; Krause, R.M.; Lane, B.W.; Graham, J.D. Intent to purchase a plug-in electric vehicle: A survey of early impressions in large US cites. Transp. Res. D 2013, 18, 39–45.

38 Skippon, S.M. How consumer drivers construe vehicle performance: Implications for electric vehicles. Transp. Res. F2014, 23, 15–31.

39 Klockner, C.A.; Naynm, A.; Mehmetoglu, M. Positive and negative spillover effects from electric car purchase to car use. Transp. Res. D 2013, 21, 32–38.

40 Nayum, A.; Klockner, C.A.; Prugsamatz, S. Influences of car type class and carbon dioxide emission levels on purchases of new cars: A retrospective analysis of car purchases in Norway. Transp. Res. A 2013, 48, 96–108.

41 Nayum, A.; Klockner, C.A. A comprehensive socio-psychological approach to car type choice. J. Environ. Psychol.2014, 40, 401–411.

42 Klockner, C.A. The dynamics of purchasing an electric vehicle—A prospective longitudinal study of the decision-making process. Transp. Res. F 2014, 24, 103–116.

43 Kang, M.J.; Park, H. Impact of experience on government policy toward acceptance of hydrogen fuel cell vehicles in Korea. Energy Policy 2011, 39, 3465–3475.

44 Tarigan, A.K.M.; Bayer, S.B. Temporal change analysis of public attitude, knowledge and acceptance of hydrogen vehicles in Greater Stavanger, 2006–2009. Renew. Sustain. Energy Rev. 2012, 16, 5535–5544.

45 Tarigan, A.K.M.; Bayer, S.B.; Langhelle, O.; Thesen, G. Estimating determinants of public acceptance of hydrogen vehicles and refueling stations in greater Stavanger. Int. J. Hydrog. Energy 2012, 37, 6063–6673.

46 Zhang, Y.; Yu, Y.; Zou, B. Analyzing public awareness and acceptance of alternative fuel vehicles in China: The case of EV. Energy Policy 2011, 39, 7015–7024.

47 Ziegler, A. Individual characteristics and stated preferences for alternative energy sources and propulsion technologies in vehicles: A discrete choice analysis for Germany. Transp. Res. A 2012, 46, 1372–1385.

48 Jensen, A.F.; Cherchi, E.; Mabit, S.L. On the stability of preferences and attitudes before and after experiencing an electric vehicle. Transp. Res. D 2013, 25, 24–32.

49 Chandra, A.; Gulati, S.; Kandlikar, M. Green drivers or free riders: An analysis of tax rebates for hybrid vehicles. J. Environ. Econ. Manag. 2010, 60, 78–93.

50 Sallee, J. Tax Credits and the Market for Hybrid Vehicles; Working paper; University of Michigan: Ann Arbor, MI, USA, 2007.

51 Berensteanuand, A.; Li, S. Gasoline price, government support and the demand for hybrid vehicles. Int. Econ. Rev.2011, 52, 161–182.

52 Chiu, Y.C.; Tzeng, G.H. The market acceptance of electric motorcycles in Taiwan experience through a stated preference analysis. Transp. Res. D 1999, 4, 127–146.

53 Can Norway's Electric Car Success Be a Model for World? The Nordic Page, Norway, 7 June 2015, Oslo. Available online: http://www.tnp.no/norway/economy/4263-can-norway-electric-car-success-be-a-model-for-worldij (accessed on 30 June 2015).

54 Diamond, D. The impact of government incentives for hybrid-electric vehicles: Evidence from US states. Energy Policy2009, 37, 972–983.

55 Wouk, V. Hybrids: Then and now. IEEE Spectr. 1995, 2, 16–21.

56 Peters, A.; Dutschke, E. How do consumers perceive electric vehicles? A comparison of German consumer groups. J. Environ. Policy Plan. 2014, 16, 359–377.

57 MacauNews. Macau to Have First Environmental Planning until 2020. MacauNews, 1 April 2011. Available online: http://www.macaunews.com.mo/index.php?option=com_contentandtask=viewandid=1248andItemid=4 (accessed on 30 January 2015).

58 Situ, L. Electric vehicle development: The past, present and future. In Proceedings of 3rd Conference on Power Electronics Systems and Applications, Digital Reference: K210509135, Hong Kong, China, 20–22 May 2009.

59 DESC. Macau Census and Statistics Department. Available online: http://www.dsec.gov.mo/ (accessed on 30 January 2015).

60 Ching, T.W. Road testing of electric vehicles in Macau. In Proceedings of the 24th Canadian Conference on Electrical and Computer Engineering (CCECE), Niagara Falls, ON, Canada, 8–11 May 2011.

61 MacauDaily Times. 'Green' Vehicles Must Be the Rule and Not the Exception, Paulo Barbosa. MacauDaily Times INSIGHT, 2 April

2012. Available online: http://www.macaudailytimes.com. mo/opinion/34954-INSIGHT-Green-vehicles-must-the-rule-and-not-the-exception.html (accessed on 30 January 2015).

62 Bento, A.M.; Goulder, L.H.; Jacobsen, M.R.; von Haefen, R.H. Distributional and efficiency impacts of increased U.S. gasoline taxes. Am. Econ. Rev. 2009, 99, 667–699.

63 Klier, T.; Linn, J. The price of gasoline and new vehicle fuel economy: Evidence from monthly sales data. Am. Econ. J. Econ. Policy 2010, 2, 134–153.

64 West, S. The Effect of Gasoline Prices and the Demand for Sport Utility Vehicles; Working paper; Macalaster College: St Paul, MN, USA, 2007.

65 Daziano, R.A.; Chiew, E. On the effect of the prior of Bayes estimators of the willingness to pay for electric-vehicle driving range. Transp. Res. D 2013, 21, 7–13.

66 Fransson, N.; Garling, T. Environmental concern: Conceptual definitions, measurement methods and research findings. J. Environ. Psychol. 1999, 19, 369–382.

67 Dunlap, R.E.; Jones, R.E. Environmental concern: Conceptual and measurement issues. In Handbook of Environmental Sociology; Michelson, W., Ed.; Greenwood: Westport, CT, USA, 2002; pp. 482–524.

68 Dunlap, R.E.; Jones, R.E. Environmental attitudes and values. In Encyclopedia of Psychological Assessment; Fernández-Ballesteros, R., Ed.; Sage: London, UK, 2003; Volume 1, pp. 364–369.

69 Poortinga, W.; Steg, L.; Vlek, C. Values, environmental concern, and environmental behavior: A study into household energy use. Environ. Behav. 2004, 36, 70–93.

70 Kahn, M.E. Do greens drive hummers or hybrids? Environmental ideology as a determinant of consumer choice. J. Environ. Econ. Manag. 2007, 54, 129–145.

71 Irwin, A.; Wynne, B. Misunderstanding Science? The Public Reconstruction of Science and Technology; Cambridge University Press: Cambridge, UK, 2004.

72 Flynn, R.; Bellaby, P. Risk and the Public Acceptance of New Technologies; Palgrave Macmillan: London, UK, 2007.

73 Wang, C.X.; Gonzalez, J.A. Assessing feasibility of electric buses in small and medium-sized communities. Int. J. Sustain. Transp. 2013, 7, 431–448.

74 Ing, A. Public acceptance of electric vehicles in Toronto. In Proceedings of the 55th Annual Meeting of the ISSS, Hull, UK, 17–22 July 2011.

75 Viardot, E. Successful Marketing Strategy for High.-Tech. Firms, 2nd ed.; Artech House: Boston, MA, USA, 1998.

76 Weigel, R.H. Environmental attitudes and the prediction of behavior. In Environmental Psychology: Directions and Perspectives; Feimer, N.R., Geller, E.S., Eds.; Praeger: New York, NY, USA, 1983; pp. 257–287.

77 Ajzen, I. Attitude structure and behaviour. In Attitude Structure and Function; Pratkanis, A.R., Breckler, S.J., Greenwald, A.G., Eds.; Erlbaum: Hillsdale, NJ, USA, 1989; pp. 241–274.

78 Sjoberg, L. Global change and human action: Psychological perspectives. Int. Soc. Sci. J. 1989, 121, 414–432.

79 Takala, M. Environmental awareness and human activity. Int. J. Psychol. 1991, 26, 585–597.

80 Liu, R.R.; Golovitcher, I.M. Energy-efficient operation of rail vehicles. Transp. Res. A 2003, 37, 917–932.

81 Department for Transport. Ultra-low carbon vehicles in the UK. Available online: http://www.berr.gov.uk/files/file51017.pdf (accessed on 30 January 2015).

82 Zeithaml, V.A.; Berry, L.L.; Parasuraman, A. The behavioral consequences of service quality. J. Mark. 1996, 60, 31–46.

83 Benter, P.M.; Chou, C.P. Practical issues in structural modeling. Sociol. Methods Res. 1987, 16, 78–117.

84 Worthington, R.; Whittaker, T. Scale development research. A content analysis and recommendations for best practices. Couns. Psychol. 2006, 34, 806–838.

85 Kline, R.B. Principles and Practice of Structural Equation Modeling, 3rd ed.; The Guilford Press: New York, NY, USA, 2011.

86 Straub, D.; Boudreau, M.C.; Gefen, D. Validation guidelines for IS positivist research. Commun. Assoc. Inf. Syst. 2004,13, 380–427.

87 Hair, J.F., Jr.; Black, W.C.; Babin, B.J.; Anderson, R.E. Multivariate Data Analysis, 7th ed.; Prentice-Hall Inc.: Englewood Cliffs, NJ, USA, 2010.

88 Burkholder, G.J.; Harlow, L.L. An illustration of a longitudinal cross-lagged design for larger structural equation models. Struct. Equ. Model. 2003, 10, 465–486.

89 Gefen, D.; Straub, D.; Boudreau, M.C. Structural equation modeling and regression: Guidelines for research practice.Commun. Assoc. Inf. Syst. 2000, 4, 1–78.

90 Bentler, P.M. Comparative fit indices in structural models. Psychol. Bull. 1990, 107, 238–246.

91 Browne, M.W.; Cudeck, R. Alternative ways of assessing model fit. In Testing Structural Equation Models; Bollen, K.A., Long, J.S., Eds.; Sage: Beverly Hills, CA, USA, 1993; pp. 136–162.

92 MacCallum, R.C.; Roznowski, M.; Necowitx, L.B. Model modifications in covariance structure analysis: The problem of capitalization on chance. Psychol. Bull. 1992, 111, 490–504.

93 Lebeau, K.; van Mierlo, J.; Lebeau, P.; Mairesse, O.; Macharis, C. Consumer attitudes towards battery electric vehicles: A large-scale survey. Int. J. Electr. Hybrid Veh. 2013, 5, 28–41.

94 Franke, T.; Krems, J.F. Interacting with limited mobility resources: Psychological range levels in electric vehicle use.Transp. Res. A 2013, 48, 109–122.

95 Peterson, S.B.; Michalek, J.J. Cost-effectiveness of plug-in hybrid electric vehicle battery capacity and charging infrastructure investment for reducing US gasoline consumption. Energy Policy 2013, 52, 429–438.

96 Robinson, J.; Brase, G.; Griswold, W.; Jackson, C.; Erickson, L. Business models for solar powered charging stations to develop infrastructure for electric vehicles. Sustainability 2014, 6, 7358–7387.

97 Caperello, N.D.; Kurani, K.S. Households' stories of their encounters with a plug-in hybrid electric vehicle. Environ. Behav. 2012, 44, 493–508.

98 Brown, S.; Pyke, D.; Steenhof, P. Electric vehicles: The role and importance of standards in an emerging market.Energy Policy 2010, 38, 3797–3806.

99 Lipman, T.E.; Delucchi, M.A. A retail and lifecycle cost analysis of hybrid electric vehicles. Transp. Res. D 2006, 11, 115–132.

100 Van Vliet, O.; Brouwer, A.S.; Kuramochi, T.; van den Broek, M.; Faaij, A. Energy use, cost and CO2 emissions of electric cars. J. Power Sources 2011, 196, 2298–2310.

CHAPTER 5

Probabilistic Method to Assess the Impact of Charging of Electric Vehicles on Distribution Grids

Eduardo Valsera-Naranjo [1], Andreas Sumper [1,2,3,*], Roberto Villafafila-Robles [1,2] and David Martínez-Vicente [1]

[1] Centre of Technological Innovation in Static Converters and Drives, Departament of Electrical Engineering, Universitat Politècnica de Catalunya-BarcelonaTech, College of Industrial Engineering of Barcelona, Carrer Comte d'Urgell, 187-08036 Barcelona, Spain

[2] Centre of Technological Innovation in Static Converters and Drives, Departament of Electrical Engineering, Universitat Politècnica de Catalunya-BarcelonaTech, School of Industrial Engineering of Barcelona, Av. Diagonal, 647, Pl. 2. 08028 Barcelona, Spain

[3] IREC Catalonia Institute for Energy Research, Jardins de les Dones de Negre 1, 08930 Sant Adrià de Besòs, Barcelona, Spain

ABSTRACT

This paper describes a grid impact analysis of charging electric vehicles (EV) using charging curves with detailed battery modelling. A probabilistic method using Monte Carlo was applied to a typical Spanish distribution grid, also using mobility patterns of Barcelona. To carry out this analysis, firstly, an IEEE test system was adapted to a typical distribution grid configuration; secondly, the EV and its battery types were modeled taking into account the current vehicle market and the battery characteristics; and, finally, the recharge control strategies were taken into account. Once these main features were established, a statistical probabilistic model for the household electrical demand and for the EV charging parameters was determined. Finally, with these probabilistic models, the Monte Carlo analysis was performed within the established scenario in order to study the lines' and the transformers' loading levels. The results show that an accurate model for the battery gives a more precise estimation about the impact on the grid. Additionally, mobility patterns have been proved to be some of the most important key aspects for these type of studies.

KEYWORDS

electrical vehicles; distribution grids; impact study

1. INTRODUCTION

The growing interest in the introduction of fossil fuel alternatives into the transportation and the large number of projects regarding this issue show that, in the future, the presence of the electric vehicle (EV) in the electric system will be significant. Important cities in the world, such as New York, have been the scenario of reference studies about the potential impact of Plug-in Hybrid EV technology on grid operations and electricity system planning [1]. This study illustrates how the proper management of charging patterns can lead to the utilisation of the current grid with any investment in infrastructure. Other studies, like [2], conclude that in cases where Plug-in Hybrid EVs obtain the energy from electricity for 40% of their miles, penetrations up to 50% by the vehicle fleet would not require the increase of electric generation capacity under optimal dispatch rules.

Therefore, it is possible to assume a future with a strong presence of EVs [3]. Under these conditions, the use of wind, solar and renewable energies is generally a potential alternative [4,5] energy source. This situation will represent an opportunity for EVs to be part of an integrated electric system, with the ability to adapt to varying conditions by performing ancillary services like power-frequency, power factor or voltage regulation. However, an EV implementation introduces complexities of this integrated system into the grid. Depending on power level, timing, and duration of the EV connection to the grid, there could be a wide variety of impacts on grid constraints, capacity requirements, fuel types used and emissions generated [6], which must be managed in order to reach an optimal dispatch rule. A control system able to optimally manage all constraints and variables of electrical transportation will lead to moderate investments into the grid and therefore to its implementation.

Simulating the most realistic scenarios is critical in order to assess the grid impact of recharging EVs. The aim of this paper is to model, in the most detailed way, the recharge load curve of EVs using detailed battery models. The most commonly used analysis reported in literature [7,8,9,10,11,12,13] uses the constant load model, which, as a conservative approach, presents the worst case, with the disadvantage that the grid impact will be overestimated. In order to proof the grid impact analysis method using detailed battery modelling, a standardised IEEE grid has been adapted to reproduce a typical distribution grid in Barcelona, combined with mobility patterns. The steps to achieve these goals are: modelling the electrical system, establishing the rules for the probabilistic model for household electrical demands and the EV demand (due to its recharge), by implementing the previous conditions in a grid and analysing the results obtained through the Monte Carlo Method.

2. STATE OF THE ART

When studying the impact of an EV fleet recharge in a grid, several aspects have to be considered prior to starting the process. Firstly, the electric grid used in the study, taking into account the type of grid, its topology, its voltage level and daily load profile, is important. Secondly, the EV type, its battery features and the EV fleet modelling should be defined. Finally, the recharging control strategy has to be considered.

When studying the impact of EV recharge on electric grids, most of the authors [7,8,10,11] chose the distribution grid type to perform their simulations. The IEEE has standardised distribution models and available data from real grids corresponding to distribution grids.

Regarding the load profile modelling, [7] used the hourly average household load curve (available from the RELOAD data base used by the Electricity Module of the National Energy Modelling System (NEMS)). A one year period with three different day types (typical weekday, weekend and peak day) and nine load types (space cooling, space heating, water heating, cooking, cloth clothes drying, refrigeration, freezing, lighting and others) were taken into consideration. In [8], a household load profile was extracted from the measurements from [14]. These measurements were taken during over the course of 24 h in 15-minute intervals with a resolution of 15 minutes during on an arbitrary winter day. Random load profiles based on probability density functions were used in [12]. Finally, in [11], grid load profiles were extracted from from the Stockholm Office of Research and Statistics, 2009 [15] (most critical case). Regarding EV modelling, most authors used a generic EV model. However, other authors chose a real model, such as the Chevy Volt [7]. The driving ranges of electric vehicles differ amongst the different types of EVs in [7,11]: the values for the driving range of electric vehicles are between 32.7 km and 64.37 km. However, regarding the state of charge (SOC) for starting the recharge, the values established for the different types are similar, between 20% and 30% [7,8,10].

Depending on the chosen EV typology, the capacity of its batteries varies drastically. For Plug-in Hybrid EV batteries, the capacity values reported in literature are: 8 kWh [12], 8.2 kWh (2.7 h of recharging time) [11] and 11 kWh (8.8 kWh considering minimum allowable SOC) [8]. In contrast, pure EVs have a capacity of 16 kWh, which almost doubles the capacity of the studied Plug-in Hybrid EVs [7].

For the specific case of the Chevy Volt EV, the main features of its lithium-ion battery are: an energy of 16 kWh, a voltage of 320 V to

350 V, a full recharge time from 6 to 6.5 h using a 110 V outlet and an electric driving range of 64.37 km (40 miles) [7].

Considering vehicle fleet modelling, several authors have taken into account the average travelled distance, the methodology for the fleet modelling and the penetration of EVs into the grid.

The travelled distance indicates the required capacity for the battery and the energy consumed by the proposed EV. The values proposed for the travelled distance per day are 0.01 km/day to 16.1 km/day [12], 16.1 km/day to 32.2 km/day [12] or 32.7 km/day [11].

Regarding the methodology for fleet modelling, in [7], its authors estimate how many vehicles would be in a transformer based on the established EV penetration and also check the transformer loading capacity. Other authors have modelled the vehicle fleet as a single large battery which comprises the sum of all batteries [10]. And finally, other authors estimated the total energy amount for private car transportation for the entire year and then it was divided by 365 in order to obtain the energy for a single day [11].

Finally, for the EV penetration, most authors use values from 0% to 20%–30% of the EV penetration [7,8,10,12,13], although one author makes the hypothesis of reaching 100% of the EV penetration [11].

The EV recharging control strategy determines how the EV obtains energy from the grid. These strategies are based on how and when the recharge is performed. According to the reviewed literature, the recharge can be performed on one of three levels: fast, quick or slow charge. The period of the day when the recharge is performed is usually determined by the peak and off-peak times of the daily load profile.

Four strategies are studied in [7]: slow charging when the vehicles arrive at home (6 pm), slow charging of peak charging, quick-charging (240V/30A) when the vehicles arrive at home (6 pm) and quick-charging (240V/30A) of peak charging. In order to deal with possible overloads in the transformer, the author proposes controlling household loads when charging vehicles through an Advanced Metering Infrastructure (AMI) and performing a stagger charge for the EV's recharge.

In [8], two possibilities are suggested: uncoordinated and coordinated charging. Uncoordinated charging is based on the concept in which each individual vehicle starts the recharge at a random time step, while with coordinated charging, the recharge is performed by minimising the grid's power losses through an optimisation problem, using quadratic programming and dynamic programming. In [11], the concept of controlled recharges is also suggested. In the studied paper, unregulated charging establishes that people recharge their cars as soon as they get home if there are no economic incentives; while in the regulated charging through Demand Side Management, the load is shifted from peak to valley periods.

Two strategies of recharge time distribution are suggested in [10]: continuous recharge and timed recharge (22:00–6:00; 21:00–9:00). It also proposes Vehicle to Grid (V2G) for the support of the grid in peak periods.

Finally, [16] introduces the concept of recharging the EV at the workplace in a recharging infrastructure where the recharge is feasible, independent of the number of available outlets and the line's loading. Four scenarios are suggested: a NO-EV scenario where no charging strategy is required, a Plug and Play scenario where no pricing mechanisms are used and where EV users recharge their cars if available outlets exist, an enhanced workplace access scenario where recharge at work is plausible even when the workplace does not offer any outlets, and an off-peak scenario where the recharge is performed between 6am and 8pm.

3. GENERAL MODEL USED FOR THE STUDY

Evaluating the impact of EV recharge on a grid through a deterministic analysis will only lead to the analysis of a specific scenario. In situations where many possibilities are expected, the most common procedure is to analyse the most critical situation. However, the most critical situation is not always the most feasible one. The Monte Carlo method allows one to ensure that, under specific conditions, the results will be within a range of values with a predefined probability. Because the model will be under deterministic and probabilistic conditions, it is important to have a valid model for both situations.

3.1. Time Horizon

The chosen time horizon for the simulations is the time lapse of 24 h. This time horizon will be divided into one hour periods. The assumption of independence amongst different days is made; therefore, no data relationship is expected between day t−1 and day *t*.

3.2. Statistical Variables

When performing Monte Carlo simulations, the first step is to define the statistical variables. Because one of the objectives of this paper is the analysis of the influence of EV recharge, the chosen statistical variables are:

- Regular electric demand
- Demand due to EV recharge

Both statistical variables will be modelled as regular loads, each one independent from the other one. In the same way as regular loads, these two variables have been modelled as an active and a reactive power, with the result of two inputs for each statistical variable; there is, however, just one input, since active and reactive power are related through a fixed power factor which does not vary.

Both regular demand and the demand due to EV recharge are modelled as aggregated load in each node. Equation (1) shows the aggregation of EVs in each node b, for each hour h and each iteration i.

$$P_{b\_h\_i} = \sum_{j=1}^{n_{EV}} P_{EVj} \quad (1)$$

3.3. Description of the Algorithm for the Probabilistic Load Flow Implementing the Statistical Model for the Household and EV Consumption

Figure 1 shows the algorithm for the load dispatching. The algorithm is based on the Monte Carlo method. In this diagram two main parts can be observed:

- Sample generation of the statistical variables
- Execution of probabilistic load flows

In the first part of the algorithm, the values for household and EV consumption are generated. Thus, the first step in the process is to obtain the statistical models for these two consumptions. For each different type of consumption, active and reactive power will be determined for each node, hour and iteration. As a result, the active and reactive power are $P_{DD\_h\_b\_i}$ and $Q_{DD\_h\_b\_i}$ for household demand and $P_{EV\_h\_b\_i}$ $Q_{EV\_h\_b\_i}$ for the EV consumption. The subscripts *DD*and *EV* stand for a consumption related to the household demand and for the EV recharge, respectively. The subscripts *h*, *b*and *i* correspond to the obtained value for the hour *h*, at the node *b* of the grid at the current iteration *i*. The statistical model for the household consumption, *"Statistical model of* $P_{DD\_h\_b\_i}$ $Q_{DD\_h\_b\_i}$*"*, is obtained by studying Spanish load profiles over the course of one year (Section 4). The statistical model for the EV consumption *"Statistical model of* $P_{EV\_h\_b\_i}$ $Q_{EV\_h\_b\_i}$*"*, comes from a study where variables, such as urban mobility, technical EV features and the number of EVs in the scenarios, are taken into account [17] (Section 5). The next step in this sample generation is to chose the number of iterations *i* and define the number of nodes *b* which comes defined for the study grid. Once these values have been defined, the samples for the household and EV consumption can be generated through their respective statistical models previously obtained.

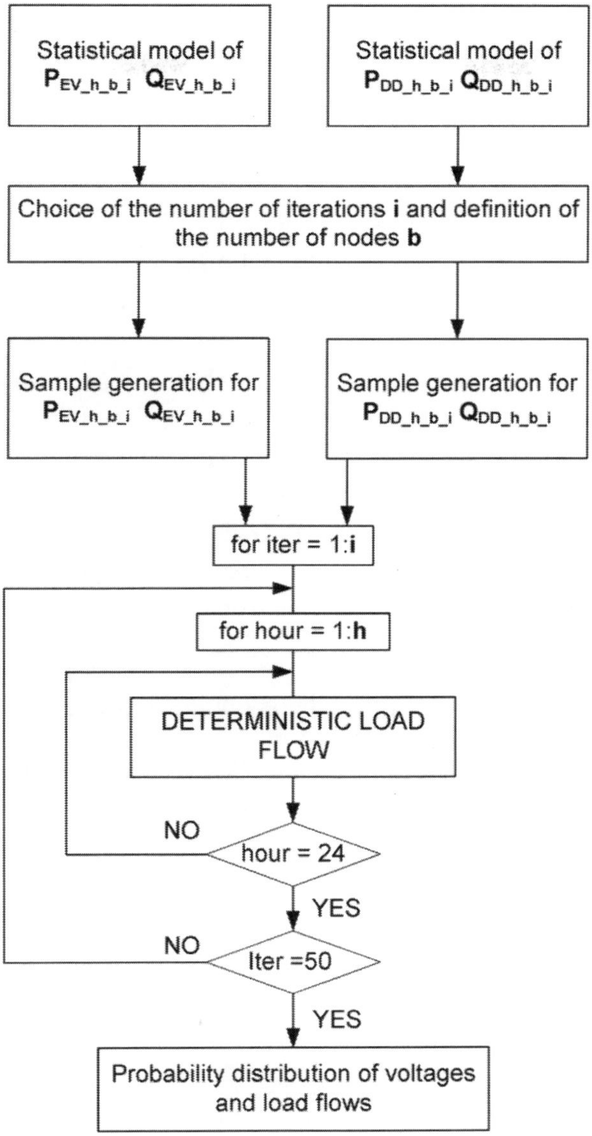

Figure 1. Flow diagram for the general procedure.

In the second part of the algorithm, the probabilistic load flow is run. It consists of an iterative process where a deterministic load flow is performed with the previously generated probabilistic values for each node, each iteration and each hour. The results of this process are the voltages of each node per load flow and iteration, and the line and transformer loadings per load flow and iteration. All these values will allow the determination of the confidence intervals and therefore the tools to compare and evaluate the different generated scenarios.

4. GENERATING THE PROBABILISTIC MODEL FOR THE GRID'S ELECTRICAL DEMAND

Electrical consumption cycles can be classified depending on its their predictable patterns. These patterns can be established either for seasons, for weekdays or for weekend days. For instance, weekdays and weekend days will have different curve shapes, even for the same week. The main factors that determine these patterns are climate and human behaviour. Therefore, it is expected that northern countries have different load patterns than from tropical countries.

A method to obtain the probabilistic model for the electrical demand is presented in this section. This model is obtained through two steps:

1. The creation of a statistical model able to generate random values according to previously defined conditions.

2. The generation of samples for the performed load flows and for each iteration.

4.1. Data Processing

In order to generate a probabilistic model for the electrical demand, several authors [18,19] have concluded that using a normal distribution is a reasonable approach. As a result, this will be the distribution used in this study.

The data used to build the statistical model has been extracted from the Spanish TSO REE (*Red Eléctrica de España*) and corresponds to Spain's electricity demand from January to September. Figure 2 depicts the daily load profiles of Spain from January to September. Each line represents a day load profile and every month has a different line colour. Different load patterns can be distinguished and since it is necessary to have an accurate model, this data is aggregated according to the similarities of the patterns.

The data corresponding to the months with a high variation of temperature was not considered because of the distortion caused in the model. Through a probability analysis, three patterns were established:

- Winter (January, February)
- Spring (April, May)
- Summer (July, August)

If the consumption profiles are grouped according to the previous classification, it is possible to get a probability distribution for each hour which follows a normal distribution. As an example, Figure 3 shows a winter labour histogram for 24 h of the power demand for Spain (Figure 3(a)) and its standard deviation (Figure 3(b)), both in MW.

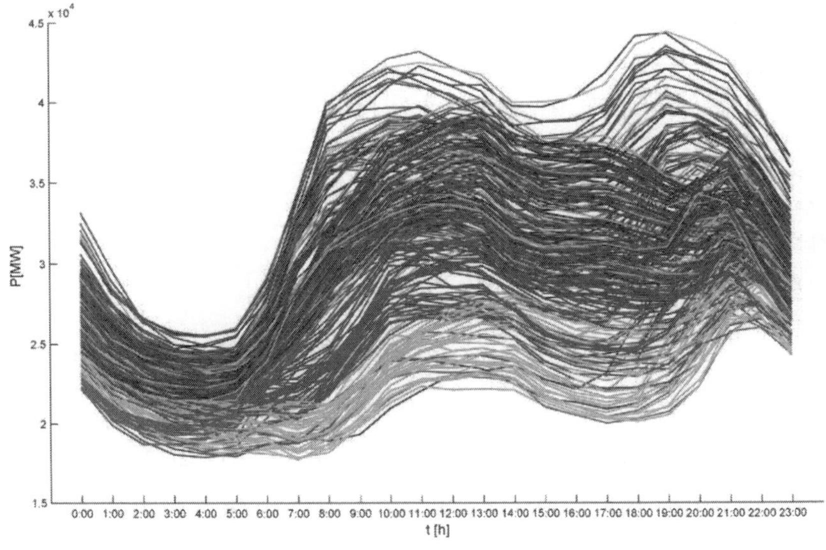

Figure 2. Daily load curves of Spain from January to September.

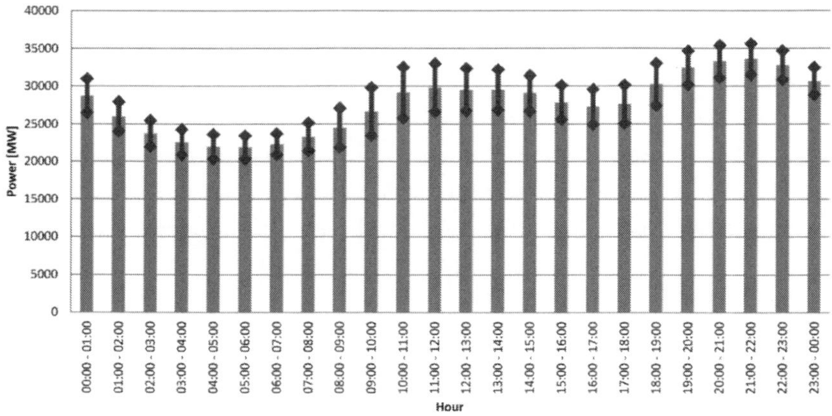

(**a**)

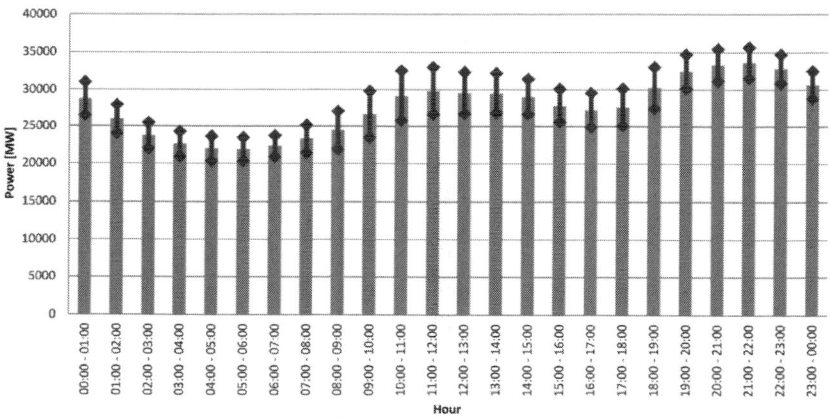

(**b**)

Figure 3. Variation of the overall demand on a labour winter day of Spain. (**a**) Mean, maximum and minimum value of the power demand; (**b**) Standard deviation of the power demand.

4.2. Adapting National Demand to a Standard Distribution Grid

Because the studied grid is a distribution grid for a specific region, it is necessary to adapt the load patterns extracted for the entire country. Therefore, the national load curve has been adapted to the study grid. The hypothesis that the placed transformers will be able to supply energy to all the types of loads (household, commercial and industrial) without exceeding the allowed loading levels when no EVs are connected to the grid is supported.

The process begins with the national load curve reduced to a unitary curve, and then a base change is performed in order to adapt the power to the capacity of each transformer station. In order to generate active and reactive power, the hypothesis of a constant power factor of 0.9 was made in order to comply with the worst case considered in the Spanish legislation [20].

5. GENERATION OF THE PROBABILISTIC MODEL FOR EV DEMAND

The objective of this section is to model the consumption generated by the EV recharge in the distribution grid in order to evaluate its impact together with the household demand. The EV consumption is difficult to model in a deterministic way because it depends on factors such as urban mobility, the EVs' technical features and the number of EVs. Therefore, it appears to be necessary to create a methodology able to take into account all these factors and also able to generate a valid statistical variable for the EVs' consumption.

The main difficulty when creating the EV statistical variable is the lack of available data. Because of this, the statistical variable which represents the EV consumption cannot be obtained from historical data, but will have to be generated from a determined number of random variables. Once these are generated, the Monte Carlo method is applied. These random variables are:

- Distribution of the EV along the grid nodes
- Connection time of the EV
- Load curve of the battery of the EV
- Energy consumed by the EV before its connection

In general terms, the two main stages of the methodology of the creation of the EVs' consumption are (Figure 4):

1. Generation of the EVs' number of connections for each hour, node and iteration
2. Transformation of the number of connections to its corresponding consumed power aggregated per node

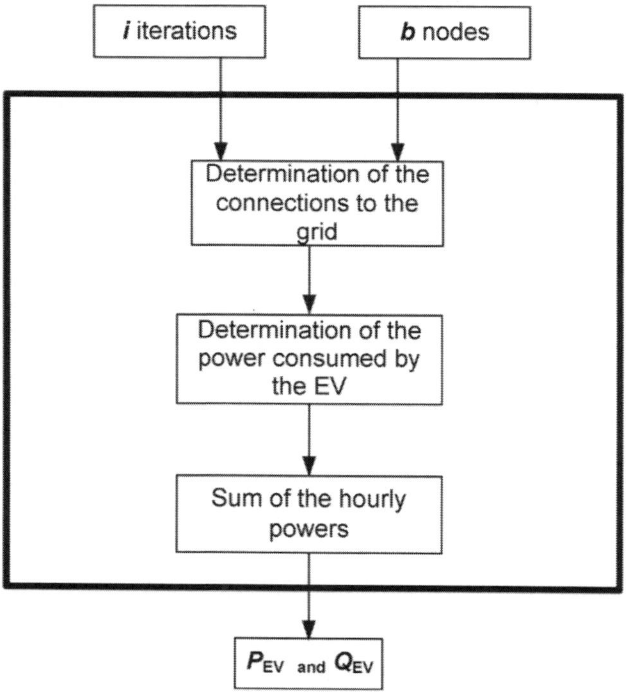

Figure 4. Flow diagram of the process "*Sample generation for EV active and reactive power (*PEV *and* QEV*)*".

5.1. Distribution of the EVs along the Grid

In order to set how the EVs are distributed along the nodes of the grid to recharge their batteries, the following variables will be taken into account:

- Number of EVs in the scenario to study
- Distribution of the connection along the nodes of the grid
- Connection time of the EV

5.1.1. Number of EV in the Studied Scenario

Since each EV can travel several times a day, the number of EVs available does not represent the number of EVs that will recharge. Thus, the decisive variable is the number of trips per node, hour and iteration (Figure 5). Because the EVs' penetration depends on the chosen scenario, the first step is to estimate the number of trips for both combustion engine vehicles and for EVs, then establish a fraction of these as electrical. For the simulations in this paper, the values for the EVs' penetration are 20% and 40% of the full number of vehicles.

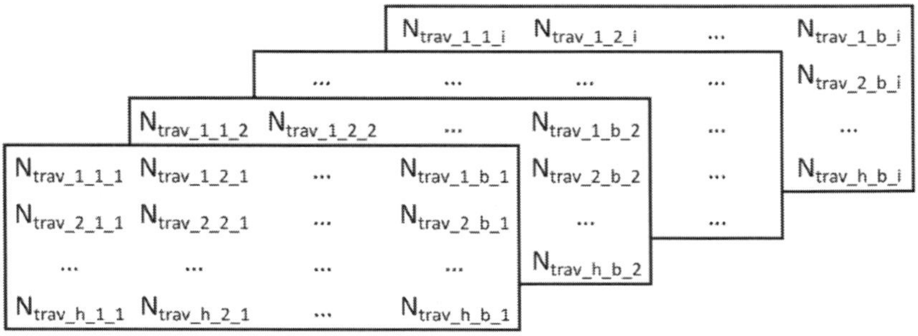

Figure 5. Matrix Ntrips.

In order to determine the number of EVs, it is necessary to set [17]:

- Number of houses (*nhouses*)
- Number of vehicles per house(*nveh/houses*)
- Mean number of trips made for each vehicle in one day (*ntrips/day*)
- Ratio of how many of these trips are made through the private vehicle (Ftrips)

With these values it is possible to determine the trips in a day *ntrips_day* through Equation (2). The number of houses is calculated by [17]

$$n_{trips\_day} = n_{houses} \cdot n_{veh/houses} \cdot n_{trips/day} \cdot F_{trips} \qquad (2)$$

Because the studied grid is an urban distribution grid, the hypothesis that 80% of the grid power is for household use is made. It is also important to consider that the contracted power in urban areas is five times the grid's capacity [17]. Finally, we assume that the contracted power for each house is 4.4 kW [17].

$$n_{house} = \frac{P_{Grid} \cdot 5 \cdot 0.8}{4.4} \qquad (3)$$

5.1.2. Distribution of the EVs along the Nodes of the Grid

The hypotheses made for the distribution of the EVs along the nodes of the grid are as follows:

1. Each node has the same probability.
2. The number of trips in the studied area remains constant. It is assumed that the number of trips going outside the studied area is the same as the number of trips going inside the studied area.

5.1.3. Connection Time of the EVs

The massive recharge of EVs during a specific time period could lead to an excessive demand of power that the system cannot deal with and to too low voltages that will break the current standards. Therefore, a realistic model is required to avoid this non-feasible situation. This model is achieved by implementing two scenarios:

1. The EV recharges its batteries after every trip.
2. The EV recharges its batteries at the end of the day when all trips are completed.

In both scenarios, the following hypotheses are assumed:

- An EV must have its battery partially or completely discharged in order to start a recharge.
- An EV can only recharge its battery if there are available outlets.
- It is assumed that EV drivers show a cautious attitude towards the possibility of ending up with a depleted battery and will therefore arrive at their destination before the battery is depleted.

The connection time of the EVs is obtained from the mobility curve of the private transport in Barcelona [21]. Because the mean trip time is 21.11 minutes on weekdays and 28.34 minutes at weekends, it is assumed that vehicles will start their recharge in the same hour when the trip is started. As a consequence, the hourly recharge curve of the EVs will have the same pattern as the mobility curve for private transport (Figure 6).

Although the mobility patterns in the entire area of Barcelona are known, it is necessary to adapt these to our area of study. Equation (4) shows how to adapt the number of travels to the proposed study grid with a specific power (*Ntrips_hour*), where ntrips'_day are all trips in the metropolitan area of Barcelona during the day, *ntrips_day* are all trips in our system (with a specific power) during the day and *Ntrips'_hour* are trips in Barcelona for the hour *h*.

$$N_{trips\_hour} = \frac{n_{trips\_day}}{n_{trips'\_day}} \cdot N_{trips'\_hour} = \begin{bmatrix} n_{trips\_1} \\ n_{trips\_2} \\ \vdots \\ n_{trips\_h} \end{bmatrix} \tag{4}$$

In order to model the number of vehicles recharging at a specific time of the day as a non-deterministic variable, it can be assumed that the arrivals will follow a Poisson distribution [22]. The Poisson distribution establishes the probability that a number *k* of events could happen in a determined period of time if they occur with a known frequency and are independent of the time of the last event. In some situations, in which the expected value is bigger than 5, it is possible to transform the Poisson distribution into a normal function [23] [Equation (5)], where $N\mu\_trips\_h$ stands for the vector with average values and $N\sigma\_trips\_h$ stands for the vector with the deviations for each hour. Then, the values can be determined for each hour, node and iteration (Figure 5).

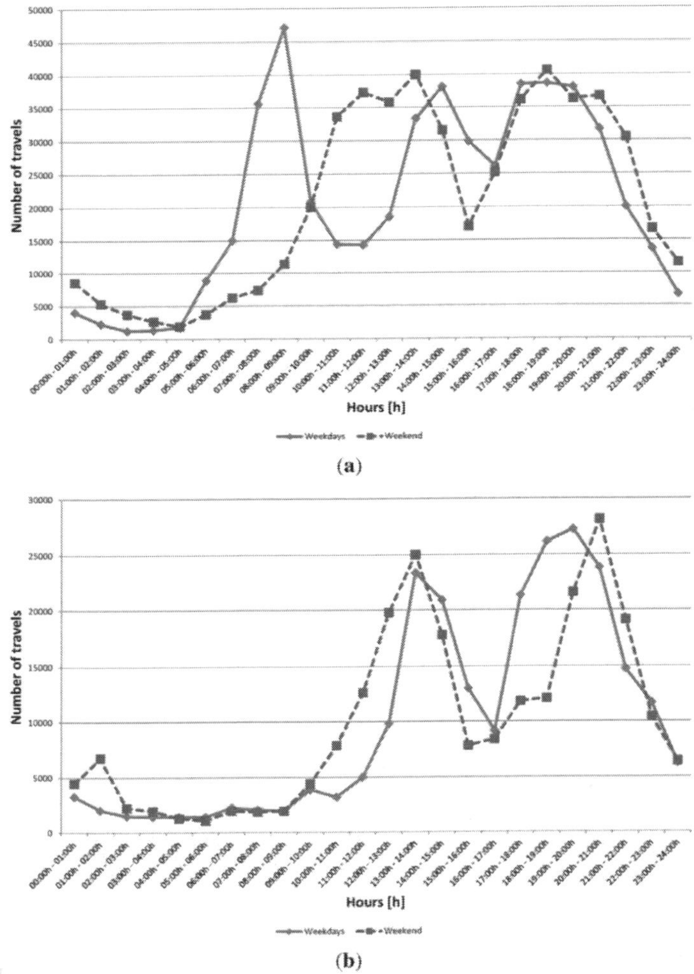

Figure 6. Mobility curves of the metropolitan area of Barcelona. (**a**) Mobility curve in thousands of travels; (**b**) Mobility curve of the travels back to home in thousands of travels.

$$N_{\mu\_trips\_h} = \begin{bmatrix} n_{trips\_1} \\ n_{trips\_2} \\ \vdots \\ n_{trips\_h} \end{bmatrix} \quad and \quad N_{\sigma\_trips\_h} = \begin{bmatrix} \sqrt{n_{trips\_1}} \\ \sqrt{n_{trips\_2}} \\ \vdots \\ \sqrt{n_{trips\_h}} \end{bmatrix} \tag{5}$$

5.2. Consumed Power Associated with EV Consumption

Once the number of trips has been determined, it is possible to associate each travel with an EV's recharge. In order to define each recharge, it is necessary to have previously defined the duration of the recharge, the amount of energy to be recharged and the battery's load curve. In summary, the analogy of the matrix active power of the EV's recharge for each hour, node and iteration will be obtained from the matrix number of trips.

5.2.1. Consumed Power by EV Recharge, *PEV_h_b_i*

Figure 7 describes the process of creating the vector for the consumed power associated with the EV's consumption. As can be seen from the figure, the needed inputs are: the *Number of trips* matrix (ntrips_h_b_i), the probability distribution of the consumed energy and the probability distribution of the battery's capacity.

The *Number of trips* matrix (*ntrips_h_b_i*) establishes the number of vehicles recharging for each hour, node and iteration (its obtainment process is described in Section 5.1).

It is important to highlight that even if two vehicles have travelled the same distance using the same battery, the process of recharging the battery does not necessarily have to be the same, since it directly depends on its SOC. Therefore, a statistical model for the total energy to be recharged has been generated from a list of batteries (Table 1). The total energy to be recharged is defined by Equation (6), where Ce is the specific consumption. In this case it is 0.158 kWh/km, obtained as the mean from Table 1, and D is the travelled distance which follows a log normal distribution with a shape value of 1.929 and a scale factor of 1.508 [17,21].

$$E_{cons\_EV} = C_e \cdot D \tag{6}$$

The type of battery which defines the load curve was also generated from (Table 1). A log normal function was created from the capacities to be generated for random batteries; and depending on the value, a known load curve from an existent model was assigned, as Figure 8 shows. The abbreviations in Figure 8 stand for location parameter (Loc), scale parameter (Scale) and sample size (N) of the log normal function obtained.

Once the number of the EV's trips is known, the EV's consumed energy per trip as well as the type of battery per trip are then determined. With these two variables, the power consumed during the EV recharge (PEV_h_b_i) is calculated. Additionally, this algorithm also considers the fact that the EV's recharge could last for more than one hour in the*Accumulated EV's power* block. As a result, the EV's power that would be in each node would be the power of the EVs that have arrived at the node b and the power of those which are still recharging because they had arrived during the previous hours.

Following this algorithm, the EV's power to recharge in each hour, node and iteration (PEV_h_b_i) can be obtained. It is important to note that this algorithm takes into account the random characteristics of the number of EV's that would recharge in each node, the consumed energy for each EV during its trip and the type of battery that the EV would have.

Finally, another conclusion that can be drawn from this process is that the duration of the connection time is a random variable which does not depend on the travel time. The connection time for each EV depends directly on the travelled distance D (which follows a log normal distribution) and the specific consumption Ce which depends on the type of battery (which also is chosen by following a log normal function).

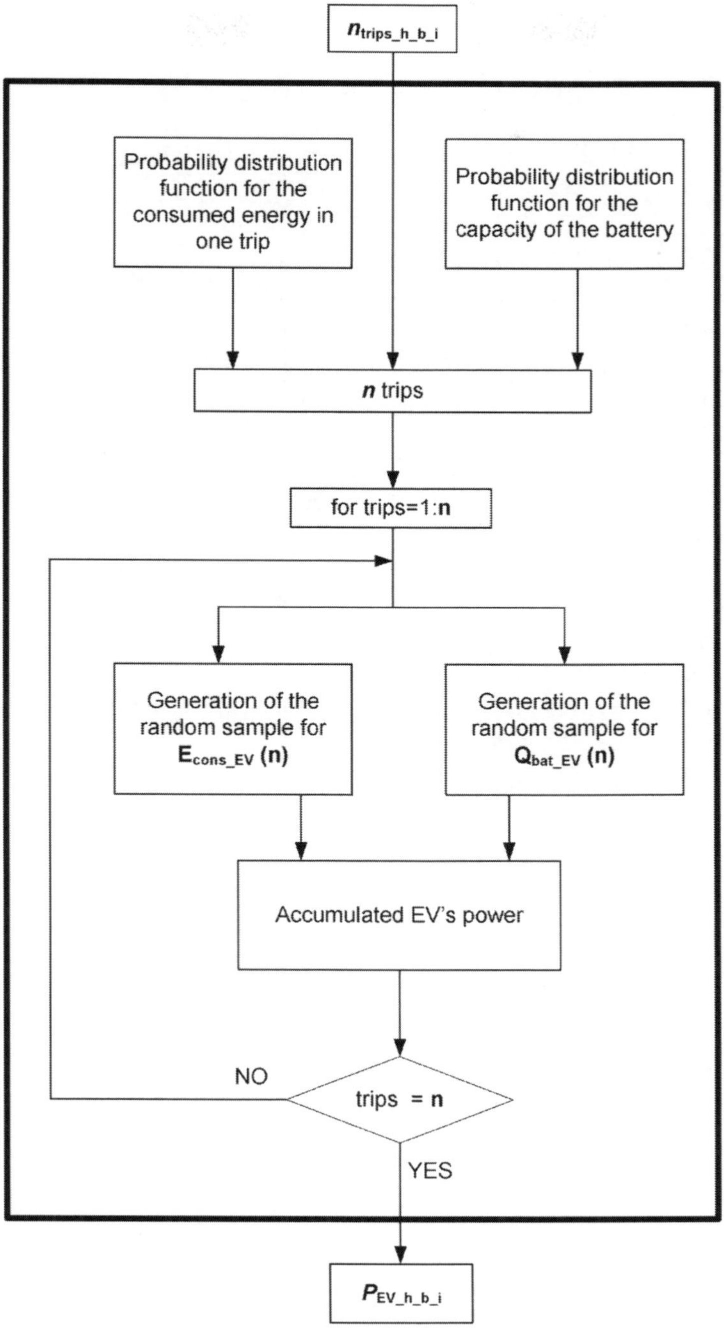

Figure 7. Flow diagram of the consumed power by the EV's recharge.

Table 1. EV and Plug-in Hybrid EV Models, Battery Type and Capacity, and Consumption.

Electric Vehicle (EV)	Capacity [kWh]	Consumption [kWh/km]	Technology
BYD F3DM	16.5	0.160	Li-Ion
BYD e6 (75kW)	59.4	0.160	Li-Ion
BYD e6 (200kW)	59.4	0.175	Li-Ion
CHANA BENNI	9.0	–	AGM
Micro-Vett Fiat 500	22.0	–	Li-Ion
Mitsubishi i MiEV	16.5	0.100	Li-Ion
Smart electric drive Coupe	14.5	0.122	Li-Ion
Subaru Estella	9.0	–	Li-Ion
Tata Indica Vista EV	26.5	0.133	Li-Ion
SEAT Leon Twin Drive Ecomotive	12.0	0.240	Li-Ion
Zytel Gorila	10.8	0.150	Lead-Acid
Opel Ampera	16.5	0.133	Li-Ion
REVA NXR	13.5	0.090	Li-Ion
Micro-Vett Fiorino M1-Fi(LC-EG)-Li	13.5	–	Li-Ion
Micro-Vett Fiorino M1-Fi(HC-Eg)-Li(S)	21.0	0.240	Li-Ion
Micro-Vett Fiorino M1-Qu(HC-Eg)-Li(S)	21.0	0.240	Li-Ion
Micro-Vett Fiorino M1-Qu(HC-Eg)-Li(L)	32.0	0.240	Li-Ion
Smart electric drive Cabrio	14.5	0.122	Li-Ion
Think City 2010	83.0	0.144	NiNa
Peugeot ION	16.5	0.125	Li-Ion
Tesla Roadster	51.5	0.231	Li-Ion
TOYOTA Prius Plug-In Hybrid	5.0	0.062	Li-Ion
Citroen C-Zero	16.5	–	Li-Ion
Mini-E	35.0	0.130	Li-Ion
Volkswagen Golf	26.5	–	Li-Ion
Nissan Leaf	24.0	–	Li-Ion
Chevrolet Volt	16.5	–	Li-Ion
Brusa Spyder	16.0	–	Li-Ion
Phoenix SUT	35.0	–	Li-Ion
Lumeneo Smera	10.0	–	Li-Ion

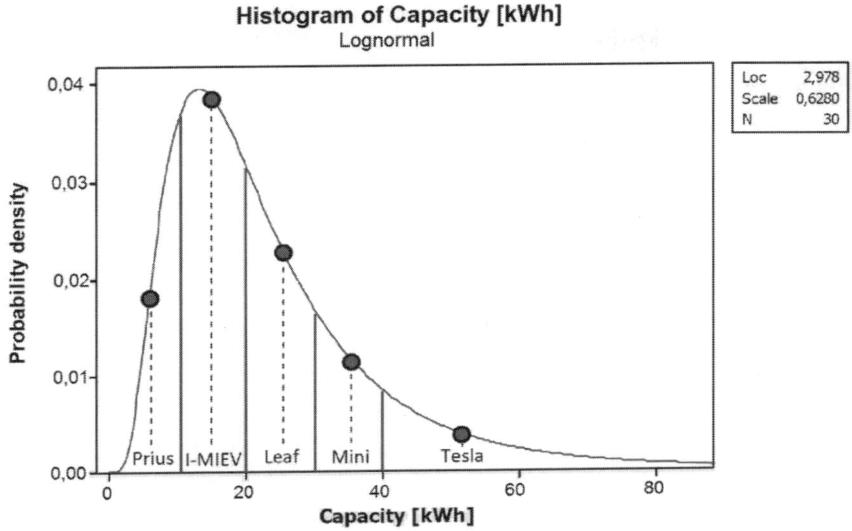

Figure 8. Distribution function of the batteries capacity.

6. STUDY GRID

In order to evaluate the impact caused by the EV on the grid, it is important to choose the scenario for the simulations, which basically consists of choosing a grid to be used. The type of grid chosen will define the accuracy of the results; thus, the more realistic the grid is, the more reliable the results will be.

When choosing the grid for the impact study, four aspects are important, namely: the real grid, the testing grid, medium voltage and low voltage. A real grid has the main advantage of its practically-oriented results, although it has the disadvantage of obtaining the data. On the other hand, a testing grid—not even being a real grid—has the advantage of the availability of the data and the certainty of working properly. Regarding the grid voltage level, the extra high voltage grid (transmission grid) has been dismissed because this grid type is intended to deal with high amounts of energy; therefore, the impact of the recharge of EVs will not have a clear effect. However, an MV (Medium Voltage) or LV (Low Voltage) distribution grid with a radial topology will allow a better interpretation of the results. Even though these grids do not have massive generation units, it is possible to find distributed generation units. Table 2 shows the advantages and disadvantages of the different types of grids.

Table 2. Benefits and drawbacks of performing the study with different types of grids.

Voltage	Topology	Benefits	Drawbacks
High	Meshed	Introduces the possibility of assessing the impact of large generation units. Lower error in the demand's modeling.	Due to the high short-circuit grid's power and its topology, a reduced impact on the grid is expected.
Medium	Radial	Brings the opportunity of introducing distributed generation units and assessing its impact. Lower error in the demand's modeling.	Due to the high short-circuit grid power, a reduced impact on the grid is expected.
Low	Radial	Due to its low short-circuit power, a high impact on the grid is expected.	It is not feasible to include any generation unit. Higher error in demand modeling.

6.1. Proposed Grid

In order to perform the study, the test system chosen was the IEEE 37 bus test system because of its length [24] (5.5 km) and due to the fact that it is similar to a typical distribution underground grid in Barcelona. Even through the IEEE 37 node test grid has features in common with the desired distribution grid, aspects like voltages and impedances have been adapted. Other grids were also studied, such as the IEEE 13 node and the IEEE 123 node, but were discarded because of their length.

6.2. Proposed Scenario

With regard to the previous discussion, it has been concluded that the study will be based on an IEEE 37 node standard test feeder (Figure 9), but adapting it to 25 kV and with all the necessary adaptations for its elements (Table 3, Table 4 andTable 5).

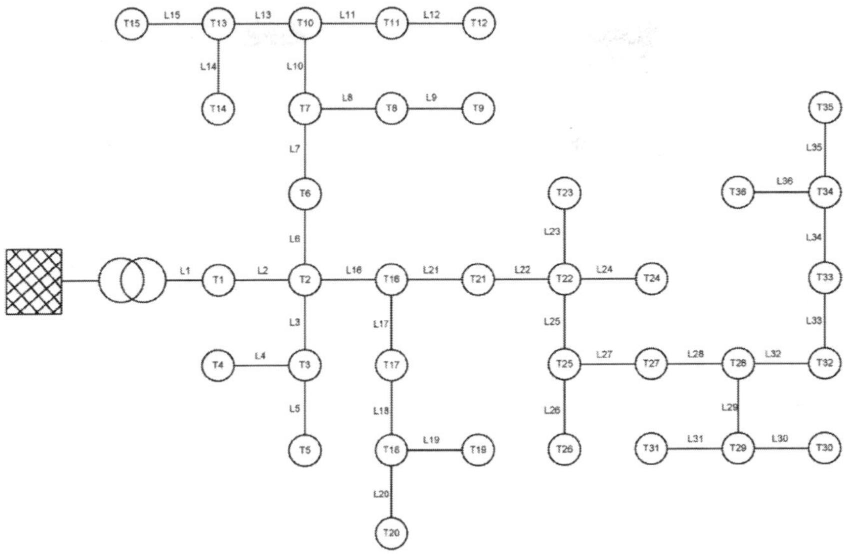

Figure 9. Studied distribution grid layout.

Table 3. Features of the proposed lines for the studied grid.

A [mm^2]	R [Ohm/km]	X_L [Ohm/km]	I_{max} [A]
240	0.125	0.116	415
400	0.0778	0.105	530

Table 4. Features of the transformers used in the studied grid.

Feature	Assigned Value
Power	160-250-400-630 kVA
Connection type: 250-400-630 kVA	Dyn11
Voltage of the HV coupling	25 kV
No-load voltage of the LV coupling	420 V
Transformer's tap positions	-5 -2.5 0 $+2.5$ $+5$ $+10$
Capacity of resisting short-circuit events in the LV side	22.2 I_{nom}

Table 5. Features of the elements used in the studied grid.

Transformer station	Number of transformers	Pnom [kVA]	Line	L [km]	S [mm²]
T0	1	25000	-	-	-
T1	1	160	L1	0.3	400
T2	3	250 + 250 + 400	L2	0.2	400
T3	1	400	L3	0.4	240
T4	2	160 + 250	L4	0.4	240
T5	1	160	L5	0.3	240
T6	1	400	L6	0.2	240
T7	1	250	L7	0.3	240
T8	2	250 + 250	L8	0.2	240
T9	1	400	L9	0.3	240
T10	1	160	L10	0.2	240
T11	2	250	L11	0.3	240
T12	1	250	L12	0.3	240
T13	2	250 + 250	L13	0.3	240
T14	1	160	L14	0.3	240
T15	2	250 + 630	L15	0.2	240
T16	2	630 + 250	L16	0.4	240
T17	2	400 + 160	L17	0.4	240
T18	1	250	L18	0.2	240
T19	1	160	L19	0.5	240
T20	2	250 + 160	L20	0.1	240
T21	1	250	L21	0.4	240
T22	2	160 + 250	L22	0.2	240
T23	2	160 + 250	L23	0.3	240
T24	1	630	L24	0.2	240
T25	2	250 + 400	L25	0.5	240
T26	2	400 + 250	L26	0.2	240
T27	1	160	L27	0.3	240
T28	1	400	L28	0.2	240
T29	1	160	L29	0.2	240
T30	2	250 + 250	L30	0.3	240
T31	1	250	L31	0.2	240
T32	2	160 + 250	L32	0.7	240
T33	1	400	L33	0.3	240
T34	2	160 + 250	L34	0.5	240
T35	2	400 + 250	L35	0.3	240
T36	1	400	L36	0.4	240

7. RESULTS WITHOUT THE ELECTRICAL DEMAND: RECHARGE CURVE OF THE EV

Several authors [7,8,10,11,12,13] performed the grid impact analysis, using a constant recharge load model. The novelty of this paper is the use of a detailed recharge curve for the impact analysis. Thus, the objective of this section is to compare the impact of the different models of batteries for the recharge of EVs, regardless of the situation of the grid. Therefore, only the demand created by the EV's recharge is considered in these simulations, which will provide a better idea of the EV's recharge effect. The different models for the EV's battery recharge are:

- **Constant Power:** The recharge is performed at constant power (Figure 10(b)).
- **Variable Power:** The recharge is performed based on the battery's recharge model (Figure 10(a)).
- **Variable model with average values for the power to recharge:** The recharge is performed based on the battery's recharge model but the chosen value for the power is the mean for the established periods (Figure 10(c)).

The simulations are performed on the chosen grid (Figure 9), with 1000 EVs during a regular weekday with 50 iterations per hour, using the software *DIgSILENT Power Factory* [25]. *DIgSILENT Power Factory* is a program to model, analyse and simulate generation, transmission, distribution and industrial grids and its interactions. This software enables as well load flow studies to be run using a scripting language called DSL. Both the recharge at the end of each trip and at the end at all day trips have been considered.

Figure 11 shows the results of these simulations. It is important to note the difference between the recharge energies in each simulation. In the case of the recharge after each trip (Figure 11(a)), it can be seen that the energy recharge directly depends on the used battery type. This difference is based on the consideration that in the same hour all vehicles will be connected to the grid, recharging at the maximum battery power. Figure 10 shows the three model types for the battery recharge of 1650 kWh The blue colour stands for the real energy required and red stands for the additional energy assumed in this simulation. Several studies from [17,21] demonstrate that the majority of the population consume less than 4 kWh during each trip.

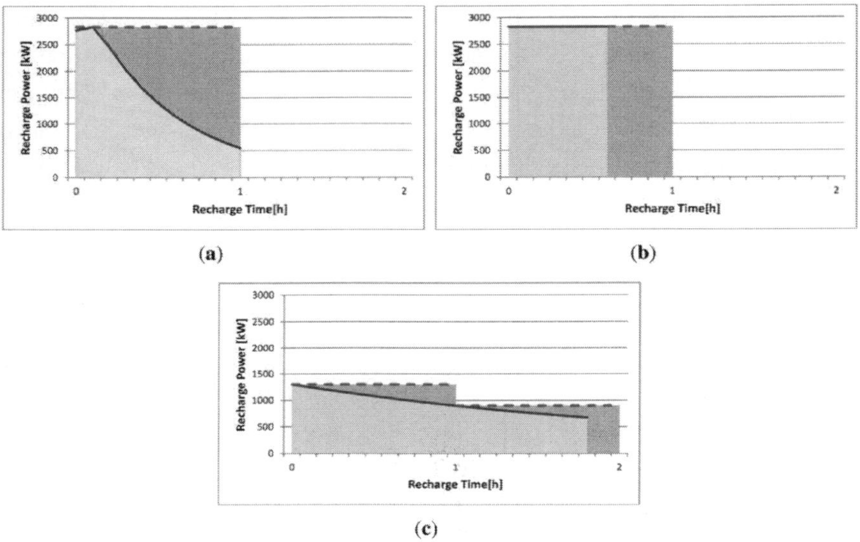

Figure 10. Recharge profile of the recharging process of three different types of batteries. (**a**) Recharge curve of a Toyota Prius's battery; (**b**) Recharge curve of a battery at constant power; (**c**) Recharge curve of a Nissan Leaf's battery.

Accordingly, battery recharges would take place in the exponential part of the curve. Therefore, in order to achieve a more precise model and not include this unrealistic extra energy, it would be necessary to measure in periods smaller than one hour when the recharge takes place in the exponential part of the model's curve.

For the case of the recharge at the end of all day trips (Figure 11(b)), the error is much smaller, since recharges are deeper and take place in the bulk part of the curve of the recharge model where minimising the error due to the exponential part of the curve.

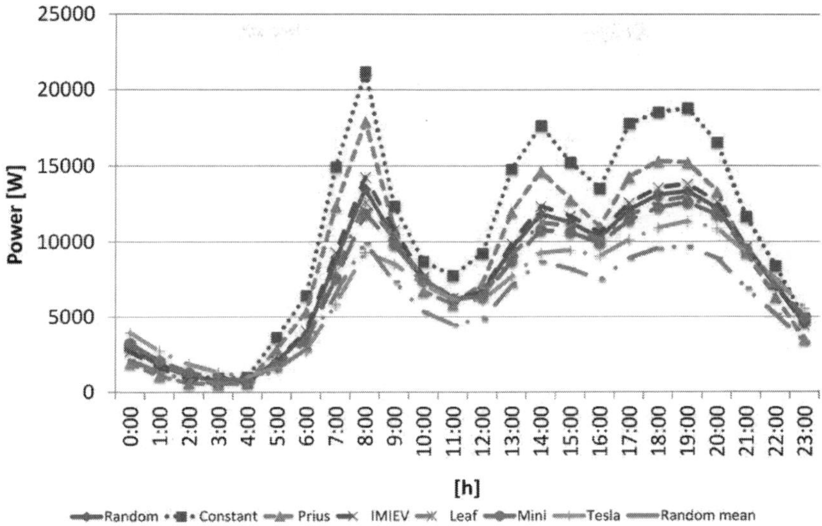

(**a**)

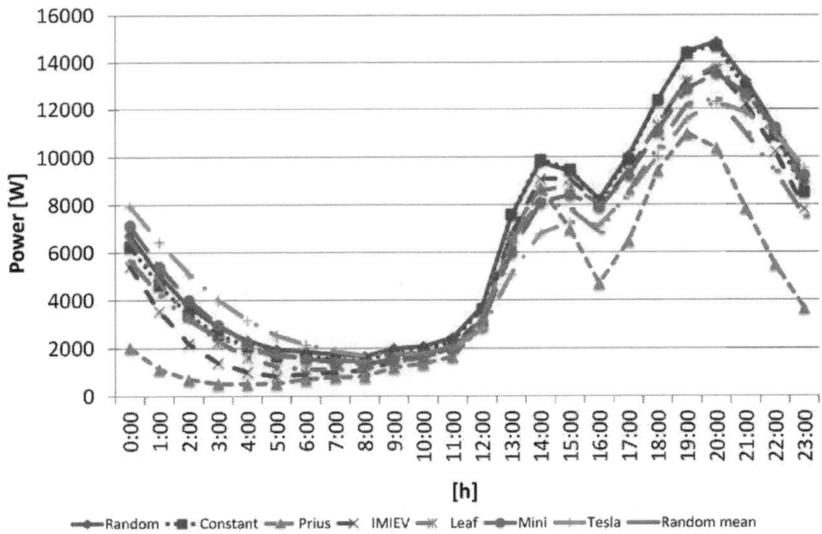

(**b**)

Figure 11. Recharge profile of 1000 EVs distributed along a 53 node low voltage grid [W]. (**a**) Recharge after each trip; (**b**) Recharge at the end of all day trips.

8. RESULTS WITH ELECTRICAL DEMAND:IMPACT INTO THE DISTRIBUTION GRID

The EV's impact on the study grid (Figure 9) is studied in this section. Prior to the simulations, it had been studied which results can provide a better understanding of the impact of EVs. Because of the topology of the grid and the household load profiles, it has been observed [17] that the greatest impact is in Line 1 (L1), and because the difference is not significant in the transformers' loadings, the value will be the mean of them all (Equation (7)).

$$L_{transf} = \frac{\sum_{i=1}^{N_{transf}} L_{transf\_i}}{N_{transf}} \tag{7}$$

8.1. Comparison of the Constant Power Battery's Recharge Model and Variable Model with Mean Values for the Power to Recharge

The results of the Monte Carlo simulation are depicted below with 0%, 20% and 40% of EV penetration for the most unsuitable scenario (the one with the highest household consumption, which in this case is the "*winter weekday*" case), using the constant power model (Figure 10(a)) and the variable model with mean values (Figure 10(c)).

As Figure 12(a) shows, the maximum load increase occurs in Line 1 (L1) for a penetration of 40% with the constant power model. This maximum load increase takes place when household demand reaches its maximum (20:00), taking the loading level from 60 % (with no EV) to 74%. In the case of the mean value model, the loading level in the same hour is 66%, which corresponds with the expected accuracy of this model (Section 7). It is important to point out that with a 40% EV penetration, the grid has already experienced a violation of the admissible level of 60% at 8:00; with the constant power model and it reaches 68% and with the mean value model it reaches 63%.

Taking into account the conclusions extracted from Section 7, it can be seen that for the recharge at the end of all day trips, the differences between the batteries' recharge models are not as significant as in the previous scenario (Figure 13); the maximum difference between the models is 1%.

Figure 13 shows the loading level for the transformers. Comparing the loading levels of the transformers with the line loading levels, it can be observed that the load pattern remains the same although the transformer case reaches higher loading levels.

8.2. Extreme Scenario: EV Penetration of 100%

The purpose of this section is to study the effect of a 100% EV penetration in the grid. Two situations will be considered: recharging after every trip and recharging at the end of the day. Results will be compared with the 0% EV penetration case.

Figure 14 depicts the loading profile of L1 and the transformer when the recharge is performed after every trip. As it can be seen in the Figure 14(a), lines do not overload their maximum value. Nevertheless, depending on the country's legislation, these values can be considered not suitable under normal operating conditions. Regarding the transformer's loading level (Figure 14(b)), it can be seen that overloading is produced all day long except during the off-peak time period.

The EV's recharging after all the trips of the days is shown in Figure 15. For the case of the line's loading (Figure 15(a)), as in the previous case, it does not exceed 100% of its capacity but it reaches higher values (the hour when this maximum value is achieved is the same for both scenarios). The transformer's loading level (Figure 15(b)) presents a decrease of its loading until 17:00. At the time of the maximum loading level (20:00) the loading is approximately 10% above of the "*Recharging after every trip case*" transformer's loading level.

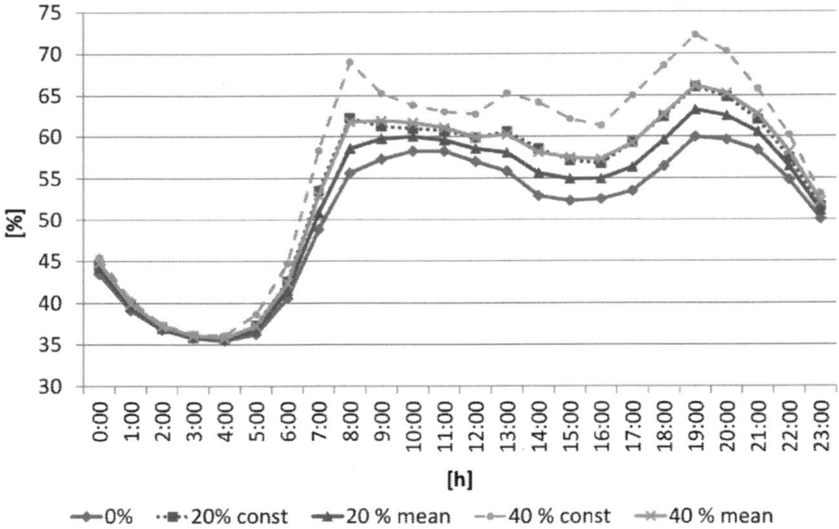

(**a**)

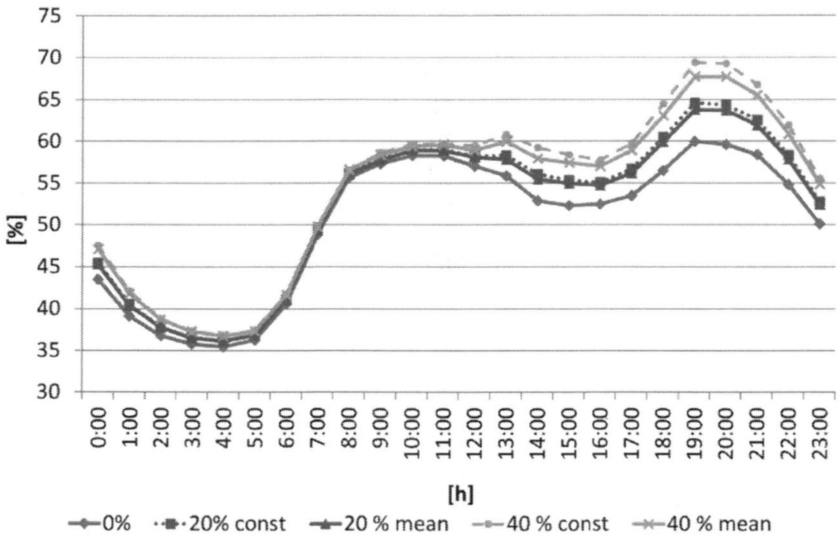

(**b**)

Figure 12. Loading profile of the Line 1 (L1) through a winter weekday. (**a**) Recharge of the battery at the end of each travel; (**b**) Recharge of the battery at the end of all day travels.

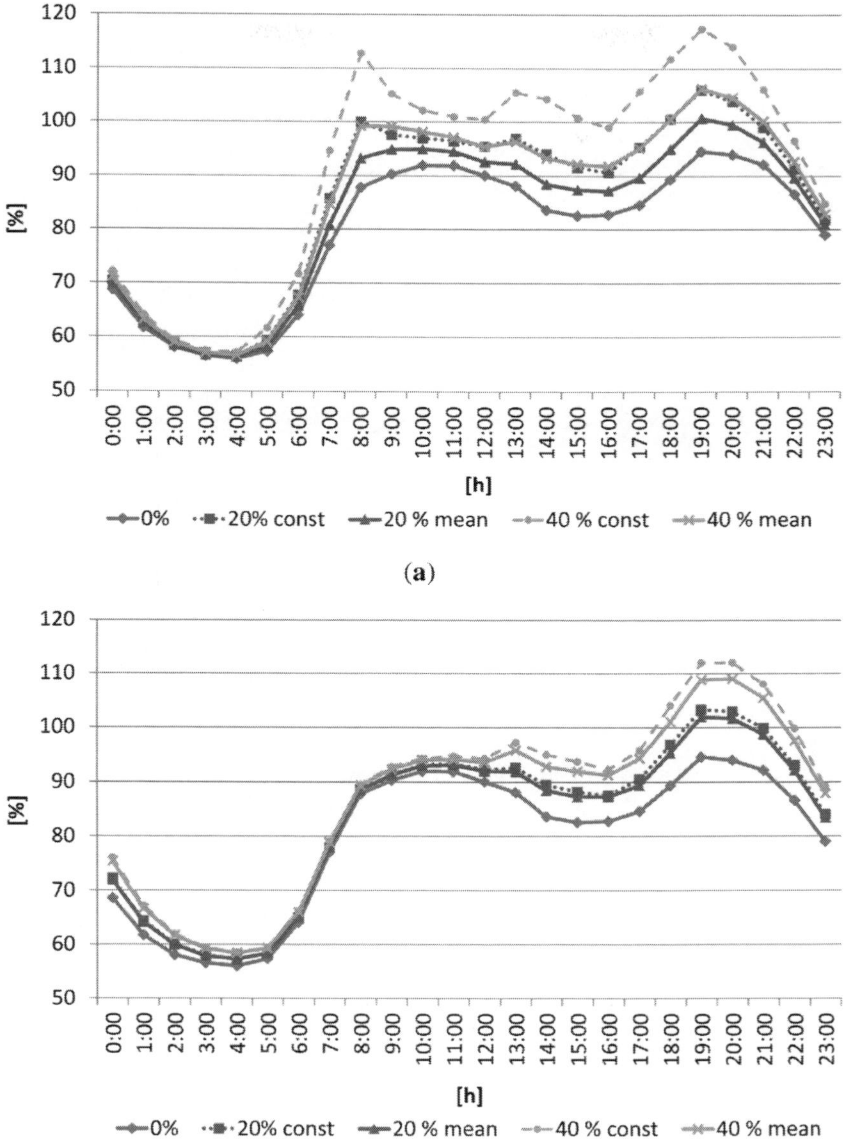

(**b**)

Figure 13. Loading profile of the transformers through a winter weekday. (**a**) Recharge of the battery at the end of each travel; (**b**) Recharge of the battery at the end of all day travels.

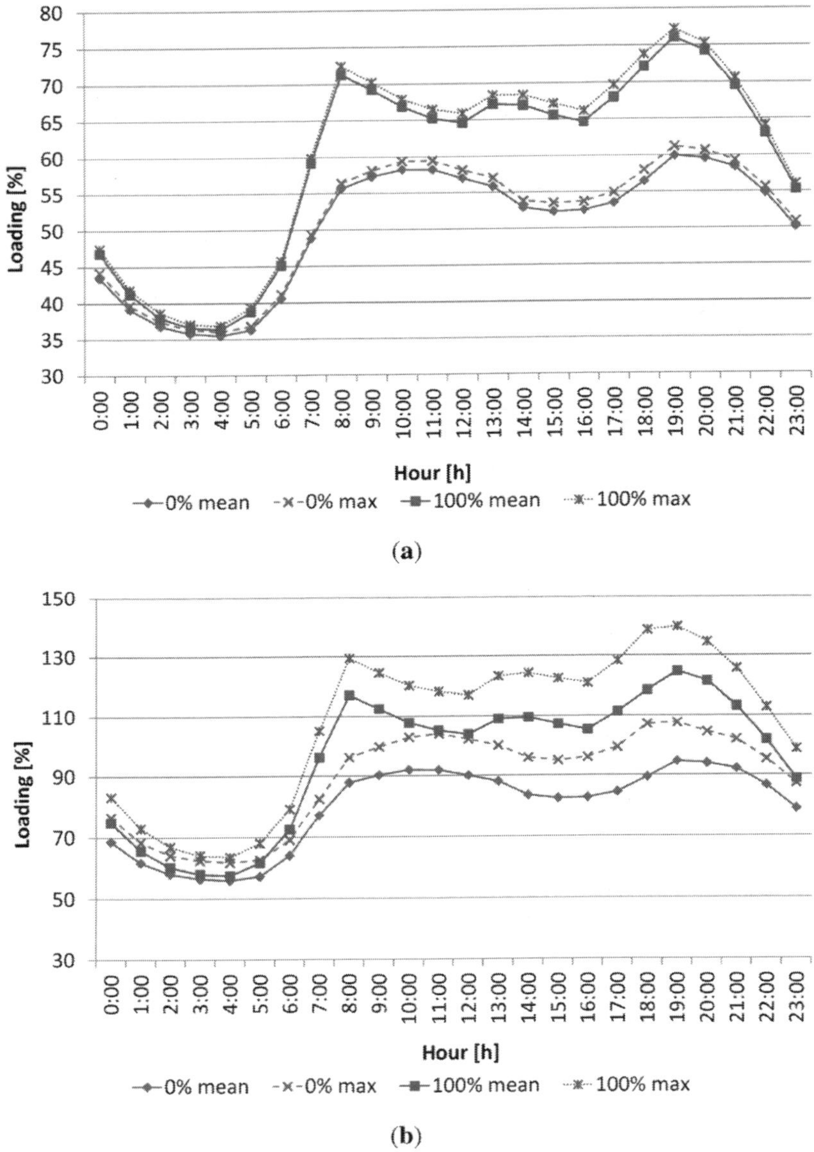

Figure 14. Grid's loading profile with an EV's penetration of 100% recharging at the end of each travel. (**a**) Loading profile of the line L1; (**b**) Loading profile of the transformers.

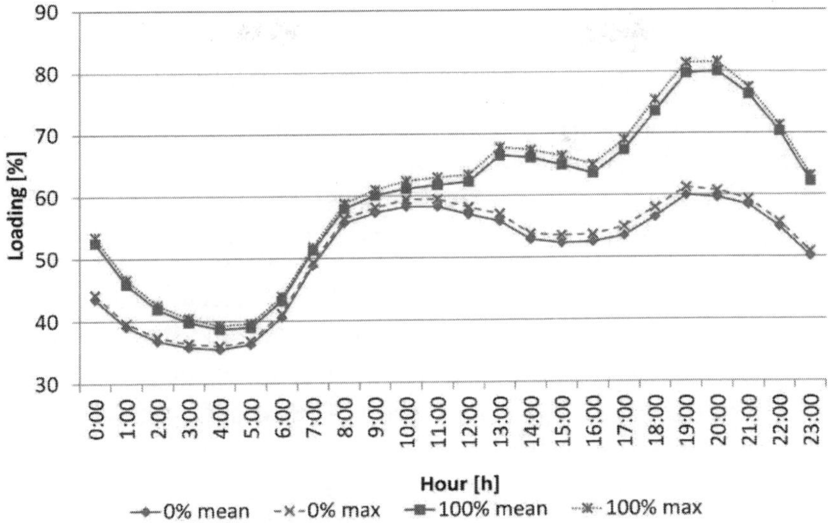

(a)

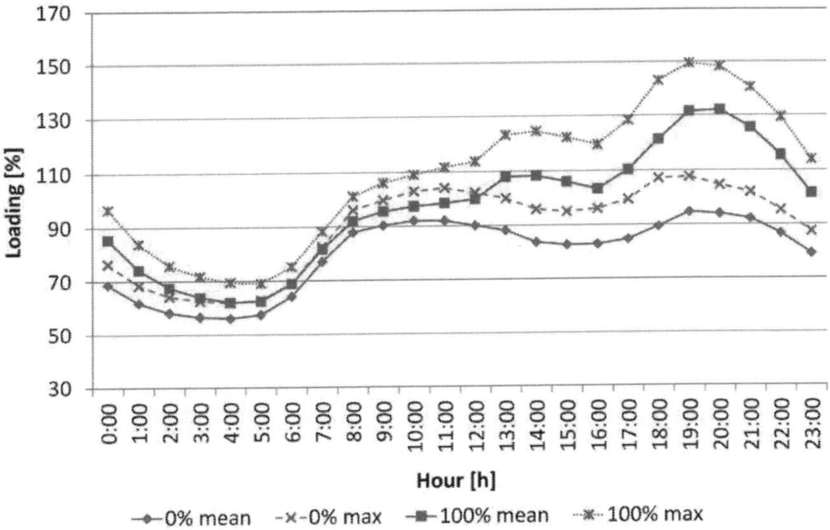

(b)

Figure 15. Grid's loading profile with an EV's penetration of 100% recharging at the end of all day travels. (**a**) Loading profile of the line L1; (**b**) Loading profile of the transformers.

9. CONCLUSIONS

The objective of this paper is to analyse the impact of the recharge of a fleet of EVs in a standard IEEE test system adopted to an underground distribution grid in Barcelona (Spain), implementing an improved model for the recharge of the batteries. In order to achieve this objective, the following tasks have been performed:

- Obtainment of the probabilistic model for electrical demand. It consists of creating a statistical model able to generate random values according to the available data of the Spanish hourly load profile (rescaled to the study grid). The seasonal patterns of the consumption were obtained in order to generate a normal distribution for each hour of each pattern. Using these normal distributions, the random values for the hourly power due to the electrical demand can be generated. The pattern chosen for the simulation is the winter pattern because the highest consumptions are experienced, which results in the most critical case.
- Battery model creation. The load curves of the different types of batteries have been implemented in order to include their behaviour in the simulations and maintain accuracy.
- Obtainment of the probabilistic model for the electrical demand due to the EV consumption. In order to get the EV consumption, several steps had to be taken:
    - Study of the mobility patterns of Barcelona and adaptation to the studied area in order to get the number of EVs that would be recharging in the area. The distribution of the number of EVs along the nodes has been done considering that each node has the same probability of recharging vehicles.
    - Determination of the consumed power of the EVs for each recharge. The consumed energy per travel is based on the specific consumption of the battery and the distance travelled. In this study the travelled distance follows a log normal distribution with a shape value of 1.929 and a scale factor of 1.508.

  - Obtainment of the type of battery of each EV. The type of battery determines the specific consumption of the previous step. In the study, a log normal function was created from the capacities to be generated for random batteries, and depending on the value, a known load curve from an existent model was assigned (Figure 8).

- One of the main interests of this study is to obtain the impact of EVs recharging in a real distribution grid situation. Thus, an IEEE test system has been adapted to an underground cable system used, like in Barcelona. The main advantage of this new test system is that it also contains mobility-related data that can be used in future studies.

- Simulation using the Monte Carlo method of the possible scenarios using the power system simulator *DIgSILENT*. Once all the consumptions per node, hour and iteration have been determined (EV's consumption and conventional demand), deterministic load flows with these values can be performed in order to get the confidence interval for the result variables.

The results demonstrate that the impact of non-constant recharge battery models on the grid is smaller than the impact of the constant one (Figure 11). This is caused by the significant inaccuracy of the energy consumed during the recharge of the constant power model, due to the following two factors:

1. The constant battery model does not take into account that during the last stage of the recharge, consumption decreases in an exponential way (constant voltage stage). This inaccuracy gets accentuated when the recharge is performed during this stage (recharge after each trip).

2. The simulation using time periods with one hour resolution results in an imprecision when the recharge lasts less than one hour. This situation results in the recharge of more energy during the simulation than required.

Despite the simulations using a non-constant battery model in order to procure the results where the impact is not the highest, during the days of the year with the highest household consumption, transformer stations could experience an overload due to the combination of the household electrical consumption and the EV recharge consumption (Figure 13). The lines will not have such severe problems (Figure 12), even though it is possible that some lines cannot remain under 60% of admissible loading power established by the utility. The same results can be observed for the critical case of the 100% of EVs penetration (Figure 14 and Figure 15).

Finally, it is essential to highlight that the variables regarding mobility aspects are some of the key factors for these kinds of studies.

ACKNOWLEDGEMENTS

The authors would like to thank the KIC-InnoEnergy project EVCity for the support received. Also thanks to Reingard Moss, Scott Maclennan and Jenny Fink for their English reviews.

REFERENCES

1 New York Independent System Operator. *Alternate Route: Electrifying the Transportation Sector*; Technical Report; New York Independent System Operator: New York, NY, USA, 2009.

2 Denholm, P.; Short, W. *An Evaluation of Utility System Impacts and Benefits of Optimally Dispatched Plug-in Hybrid Electric Vehicles*; Technical Report; National Renewable Energy Laboratory: Colorado, USA, 2006.

3 Gong, Q.; Midlam-Mohler, S.; Marano, V.; Rizzoni, G. Study of PEV charging on residential distribution transformer life. *IEEE Trans. Smart Grid* **2012**, *3*, 404–412.

4 Dickerman, L.; Harrison, J. A new car, a new grid. *IEEE Power Energy Mag.* **2010**, *8*, 55–61.

5 Gonzalez, C.; Villafafila, R.; Ramirez, R.; Sudria, A.; Sumper, A.; Chindris, M. Assess the Impact of Photovoltaic Generation

Systems on Low-Voltage Network: Software Analysis Tool Development. In Proceedings of the 9th International Conference Electrical Power Quality and Utilisation (EPQU 2007), Barcelona, Spain, 9–11 October 2007; pp. 1–6.

6 Hadley, S. *Impact of Plug-in Hybrid Vehicles on the Electric Grid*; Technical Report; Oak Ridge National Laboratory: Oak Ridge, TN, USA, 2006.

7 Shao, S.; Pipattanasomporn, M.; Rahman, S. Challenges of PHEV Penetration to the Residential Distribution Network. In Proceedings of the IEEE Power Energy Society General Meeting (PES '09.), Calgary, Canada, 26–30 July 2009; pp. 1–8.

8 Clement-Nyns, K.; Haesen, E.; Driesen, J. The impact of charging plug-in hybrid electric vehicles on a residential distribution grid. *IEEE Trans. Power Syst.* **2010**, *25*, 371–380.

9 Clement, K.; Haesen, E.; Driesen, J. Coordinated Charging of Multiple Plug-in Hybrid Electric Vehicles in Residential Distribution Grids. In Proceedings of the IEEE/PES Power Systems Conference and Exposition (PSCE '09), Seattle, WA, USA, 15–18 March 2009; pp. 1–7.

10 Acha, S.; Green, T.C.; Shah, N. Effects of optimised plug-in hybrid vehicle charging strategies on electric distribution network losses. In Proceedings of the IEEE Transmission and Distribution Conference and Exposition (2010 PES), New Orleans, LA, USA, 19–22 April 2010; pp. 1–6.

11 Karnama, A. Analysis of Integration of Plug-in Hybrid Electric Vehicles in the Distribution Grid. MA Thesis, Royal Institute of Technology (KTH), Stockholm, Sweden, 2009.

12 Taylor, J.; Maitra, A.; Alexander, M.; Brooks, D.; Duvall, M. Evaluation of the Impact of Plug-in Electric Vehicle Loading on Distribution System Operations. In Proceedings of the IEEE Power Energy Society General Meeting (PES '09.), Calgary, Canada, 26–30 July 2009; pp. 1–6.

13 Judd, S.; Overbye, T. An Evaluation of PHEV Contributions to Power System Disturbances and Economics. In Proceedings of the 40th North American Power Symposium (NAPS '08.), Calgary, Canada, 28–30 September 2008; pp. 1–8.

14 Vlaamse regulator van de elektriciteits. De verbruiksprofielen van huishoudelijke en niethuishoudelijke elektriciteitsverbruikers voor het jaar 2007. Available online: http://www.vreg.be (accessed on 15 October 2010).

15 Stockholm Office of Research and Statistics. *Statistical Year-book of Stockholm*; Stockholm Office of Research and Statistics: Stockholm, Sweden, 2009.

16 Axsen, J.; Kurani, K.S. Anticipating plug-in hybrid vehicle energy impacts in California: Constructing consumer-informed recharge profiles. *Trans. Res. Part D* **2010**, *15*, 212–219.

17 Martinez-Vicente, D. El impacto del vehiculo electrico en la red de distribucion. MA Thesis, Universitat Politècnica de Catalunya (UPC), Barcelona, Spain, 2011.

18 Seppala, A. Load Research and Load Estimation in Electricity Distribution. Ph.D. Thesis, Helsinki University of Technology, Helsinki, Finland, 1996.

19 Neimane, V. On Development Planning of Electricity Distribution Networks. Ph.D. Thesis, Royal Institute of Technology, Stockholm, Sweden, 2001.

20 Ministry of Science and Technology. *Reglamento Electrotecnico de Baja Tension (Low Voltage Electrical Standard)*; Spanish Government: Madrid, Spain, 2002.

21 *EMQ 2006 de la Regiò Metropolitana de Barcelona*; Generalitat de Catalunya —Departament de Política Territorial i Obres Públiques: Barcelona, Spain, 2006.

22 Vlachogiannis, J. Probabilistic constrained load flow considering integration of wind power generation and electric vehicles. *IEEE Trans. Power Syst.* **2009**, *24*, 1808–1817.

23 Ronald, E.W.; Raymond, H.M.; Sharon, L.M. *Probability and Statistics for Engineers and Scientists*, 9th ed.; Prentice Hall: New Jersey, NJ, USA, 2011.

24 Kersting, W.H. Radial distribution test feeders. *IEEE Trans. Power Syst.* **1991**, *6*, 975–985.

25 DIgSILENT PowerFactory. Available online: http://www.digsilent. de (accessed on 15 May 2012).

CHAPTER 6

Probabilistic Agent-Based Model of Electric Vehicle Charging Demand to Analyse the Impact on Distribution Networks

Pol Olivella-Rosell [1], Roberto Villafafila-Robles [1,2], Andreas Sumper [1,2,*] and Joan Bergas-Jané [2]

[1] Centre d'Innovació Tecnològica en Convertidors Estàtics i Accionaments (CITCEA-UPC), Departament d'Enginyeria Elèctrica, Universitat Politècnica de Catalunya, EU d'Enginyeria Tècnica Industrial de Barcelona, Carrer Comte d'Urgell, 187-08036 Barcelona, Spain

[2] Centre d'Innovació Tecnològica en Convertidors Estàtics i Accionaments (CITCEA-UPC), Departament d'Enginyeria Elèctrica, Universitat Politècnica de Catalunya, ETS d'Enginyeria Industrial de Barcelona, Av. Diagonal, 647, Pl. 2. 08028 Barcelona, Spain

ABSTRACT

Electric Vehicles (EVs) have seen significant growth in sales recently and it is not clear how power systems will support the charging of a great number of vehicles. This paper proposes a methodology which allows the aggregated EV charging demand to be determined. The methodology applied to obtain the model is based on an agent-based approach to calculate the EV charging demand in a certain area. This model simulates each EV driver to consider its EV model characteristics, mobility needs, and charging processes required to reach its destination. This methodology also permits to consider social and economic variables. Furthermore, the model is stochastic, in order to consider the random pattern of some variables. The model is applied to Barcelona's (Spain) mobility pattern and uses the 37-node IEEE test feeder adapted to common distribution grid characteristics from Barcelona. The corresponding grid impact is analyzed in terms of voltage drop and four charging strategies are compared. The case study indicates that the variability in scenarios without control is relevant, but not in scenarios with control. Moreover, the voltages do not reach the minimum voltage allowed, but the MV/LV substations could exceed their capacities. Finally, it is determined that all EVs can charge during the valley without any negative effect on the distribution grid. In conclusion, it is determined that the methodology presented allows the EV charging demand to be calculated, considering different variables, to obtain better accuracy in the results.

KEYWORDS

electric vehicles charging; agent-based modelling and simulation; distribution network; coordinated charging; load flow analysis; stochastic modelling

1. INTRODUCTION

Electric vehicles (EVs) are presented as an alternative to current internal combustion vehicles powered by fossil fuels. Increasing oil prices, greenhouse gas emissions and environmental concerns of citizens boost interest in this technology. Energy supply from power networks is required and the impact on the distribution grids in a massive EV integration scenario has to be analyzed in detail [1]. Thus, studies about EV impact on power networks are needed to ensure the viability of the systems [2–4].

The EV charging demand model should allow the analysis of possible effects of this new demand supplied in present-day power networks.

In order to do so, an EV charging model should include specific characteristics for each case, such as mobility, and it should allow one to compare different cases. Moreover, it should consider probability distribution functions (PDF) to analyze the uncertainties of possible EV charges. In addition, this model should be designed to analyze the application of control strategies and enable their comparison.

Literature proposes models to calculate the demand with respect to vehicle, charging infrastructure, mobility, and social parameters. [5–9] use different parameters such as EV model, distance, and charging process among others to determine the EV charging demand.

1.1. EV Type

From the point of view of EV charging demand, EVs main characteristics are the vehicle type: Plug-in Hybrid Electric Vehicle (PHEV) or Battery Electric Vehicle (BEV), battery capacity, battery technology, EV range and energy consumption. Amjad *et al.* [10] expose an analysis about EV design considerations. Different authors only consider PHEV [2,3,5,11–15]. Others only BEV [16–20] or a combination of both [9,21–23]. Another option is to suppose average EV models, BEV and PHEV, with average characteristics like different authors do [5,6,18]. Pang *et al.* [24] simulate only two representative EV models: Chevy Volt (PHEV) and Nissan Leaf (BEV) and Valsera *et al.* [17] simulate Mitsubishi i-MiEV (BEV) only.

Soares *et al.* [9] proposed a stochastic model with mobility variables, but the vehicle characteristics are determined by a Gaussian distribution with standard values for the capacity, energy consumption and charging power of EVs.

The majority of papers simplify the EV model selection, but the capacity and the energy consumption are significant variables to be considered. The model presented proposes using real EV models and their technical data to define the battery capacity and energy consumption of each EV model. Moreover, the probability of each EV model is based on sales forecasting [25] to decide which EV model is more probable.

1.2. Battery and Charging Process

Regarding EV batteries, there are three variables linked: capacity (kWh), range (km) and energy consumption (kWh/km). [19,26] consider the battery characteristics of real models and [15,20,27,28] consider average battery characteristics. Moreover, it is important to take into account the relation between the power consumed and the State-of-Charge (SoC). Valsera *et al.* [17] determine a relation between EV model, battery characteristics (Li-ion, 50 Ah, 16 kWh and 330 V) and its charging process.

The charging process standards of IEC 61851 [29] from Europe and SAE J1772 [30] from the USA could also change the impact in the power system. Maitra *et al.* [22] compare the impact of each SAE standard. The voltage level in Europe for slow charges is 230 V and a maximum current of 16 or 20 A. In Belgium, houses have a protection up to 20 A [31] and in Spain, the common protection is up to 16 A [17]. Valsera *et al.* use the power ratio of Mitsubishi i-MiEV when the initial SoC is 20% and the EV needs 4 h to reach 100%. Zhang *et al.* [32] use level 1 (120 V–15 or 20 A) in the studio located in the United States. To compare, Grenier *et al.* [33] use 230 V and 15 A and the study is located in New Zealand. The efficiency used in the studies is around 90%, as Collins *et al.* proposed [34] in 1983 and this assumption was recently confirmed by Shuang *et al.*[12] and Clement *et al.* [2,35].

Different authors, such as Clement *et al.* [2] and Guo *et al.* [36], use constant power profiles. On the other hand, Maitra *et al.* [22] consider variable power during the charging profiles. Qian *et al.* [37] propose a charging process model which links the power of the charger and SoC. Gao *et al.* [38] link the SoC and the charging time. Different authors use the specific EV charging profile of a real EV. For example, Qian *et al.* [37] and Lojowska *et al.* [39] use the charging profile of the Nissan Altra EV with a battery of 29 kWh, while Multin *et al.* [40] use a three-phase charging profile of Opel Meriva, which has a battery of 16 kWh.

1.3. Charging Infrastructure

Charging infrastructure parameters include the EV charging point's socket and availability to charge. The majority of works do not consider the EV infrastructure when calculating the EV charging demand. Inherent to this hypothesis is to neglect the effect of the queues at charging points by supposing there are enough charging stations, and the assumption of full compatibility between charging stations and EV connectors. Both could be reasonable in future scenarios with massive presence of EV, but could be a problem for fast chargers. García-Valle *et al.* [41] introduce the queue theory with exponential distribution function to simulate EV charging time and relate it to the maximum charging power of the EV.

1.4. Mobility

Mobility is the third key point of EV charging demand. There is a strong link between energy consumption of EV and urban mobility. For example, Keirstead *et al.* [42] reviewed the energy consumption in urban areas, including electric mobility.

Some authors employ the NHTS (National Household Travel Survey) to analyse the United States, such as [21,32,36,43–45]. In the United Kingdom, studies use NTS (National Travel Survey) and UKTUS (United Kingdom Time Use Survey), for instance [3,46,47]. In Germany, there is the MID (*Mobilität in Deutschland*) which Schroeder *et al.* [48] and Loise [20] apply. The MON (*Mobiliteitsonderzoek Nederland*) is utilised by Dutch studies, as Lojowska *et al.* [39]. The DTU *Transport, DTU. Transportvaneundersøgelsen* is used by Jull *et al.* [49] for a case study of Denmark. In the case of Spain, there are different databases, for example *Dades Bàsiques de Mobilitat 2008* for Barcelona city [17] and *MOVILIA* for the whole Spain [50].

Metz [27] makes use of the Deutsches Mobilitätspanel to simulate 1000 mobility of household profiles and this includes day and time of departure and arrival, travel distance, vehicle used, and destination. Loise [20] makes projections of EV hourly charging profiles based on MID 2008.

The present work proposes that the reason of displacement be included to determine the destination and the instant of the day to displace. Due to that, it is possible to distinguish between professional and personal mobility.

1.5. Social

There are social variables related to the EV driver profile that could influence EV charging demand as GDP. Kelly *et al.*[43] analyze the EV charging demand considering the income, age and gender of drivers as well as the location (urban or rural). Sikai *et al.* [46] use the number of members of each household and the corresponding number of vehicles based on the UKTUS database. Valsera *et al.* [8] define the number of displacements, the number of houses, and the number of vehicles per house. The proposal of the present work is to combine these three approaches of the previous work to consider social aspects to calculate the EV charging demand.

1.6. Simulation Techniques

To define the characteristics of simulations, there are different details set out by each author. The first one is the data processing, after that the emulation of parameters and lastly, the driver behavior emulation.

Considering data processing, there are different types of simulation models to emulate the EV charging demand and the most used is agent-based. This type of model considers each EV driver autonomously defining the internal (e.g., energy consumption) and external (e.g., power demand to supply EV battery) variables. The bottom-up approach simulates the system coupling all the agents of the system. Different examples of agent-based and bottom-up approach studies are [44,51,52]. On the other hand, the bottom-down approach simulates the EV driver behavior with the average parameters [17,22].

As concerns the emulation of parameters, some models use deterministic variables and others stochastic ones. The deterministic approach considers just average values of parameters and stochastic models use probability distribution functions. The Monte Carlo technique is used to simulate stochastic variables in many applications and it is also used in modeling load, EV charging demand and distributed generation to determine their variability. The majority of studies set out a deterministic approach, but some of them include stochastic variables such as [3,17,20,36,39,46,47]. Some of them use Monte Carlo techniques to simulate the total demand.

EV driver behavior also influences the EV charging demand. This parameter is linked to time of day and location for EV charging, such as public stations between trips, at charging points at work or just home charging.

Venkatesan *et al.* [53] define user profiles related to estimated behavior in the function of mobility, current electricity price and price forecasting. Waraich *et al.* [51] use microsimulation techniques to emulate the driver behavior. Galus and Waraich [54,55] use MATSim (Multi-Agent Transport Simulation) and this tool allows the creation of more than a million connections between agents in transport issues. Balmer [56] uses evolutionary algorithms; Hedegaard *et al.* [57] propose using the Balmorel program to include distribution network, district heating, optimization, taxes and geographical data.

[58] proposes including the game theory to simulate the interaction between agents and including sale of electricity with V2G service. Smith *et al.* [59] use GPS data and EV metering to calculate the energy consumption and later to optimize the battery sizing of future PHEV.

The present work proposes combining some characteristics presented in literature. The methodology presented is a bottom-up approach to process the data with stochastic variables following the Monte Carlo formulation to emulate the parameters. And the driver behavior is defined in function of the range anxiety, the mobility needs and the energy price.

1.7. Power System Impact

Possible effects on power networks caused by EVs are related to power quality or grid saturation. The majority of studies analyze the voltage drop or transformer load, like Valsera *et al.* [8,17]. Clement *et al.* [2] include Joule losses and Maitra *et al.*[22] include overloading and unbalances. Kleiwegt *et al.* [60] propose a methodology to detect overloads in the course of a year. Moreover, vehicle-to-grid possibility is analyzed in many studies such as [26,61,62]. Another possible impact on the power system is economic and this is reviewed by Dallinger *et al.* [63]. The present work analyses the distribution network in terms of the HV/MV and MV/LV transformer capacities and the voltage of each node.

1.8. Contribution

The state-of-the-art analysis defined seven subjects to be determined in the EV charging demand problem formulation:

- EV type and model: the majority of current models simplifies this aspect with one model or an averaged model to represent a group of models.
- Battery and the corresponding charging process: according to the literature review, the main difference found in literature is the charging process. The most common simplification is to consider a constant power but the appropriate way is to consider the relation between the SoC and the power consumed.
- Power infrastructure: the majority of articles consider the AC slow charging and the current limit depends in function on the country analyzed.

- Mobility: the papers which consider it try to use the public data according to the country analysed.
- Social: the majority of the papers do not consider any economic or social variables.
- Simulation technique: the majority of papers take a bottom-down deterministic approach.analyzed
- How to analyze the impact on the power system: the majority of EV charging models avoid this issue and some of them try to optimize the EV charges to reduce some negative consequences.

The objective of this paper is to define a methodology based on agents to determine EV charging demand. The main contribution of this paper is to propose a methodology based on open data and combining social, technical and economic variables to calculate the EV charging demand and then determine the effects on the distribution networks. To do so, the parameters in literature were used separately; however, this paper proposes that all of them be combined in a single model in order to obtain more precise and realistic results. Figure 1 shows the relation among the variables that are implemented in the present model. For example, EV agents have a set of constant parameters as EV model (technical), place of residence (social), GDP (economic) and others, as well as variable parameters of mobility such as distance, day of the week and others.

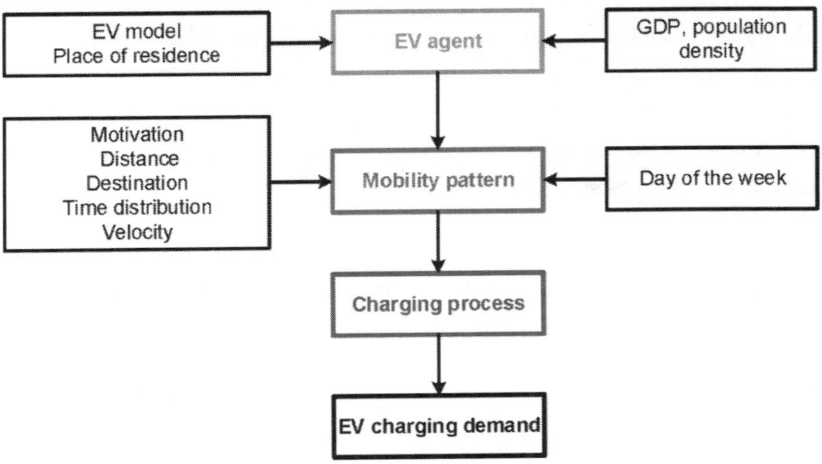

Figure 1. Basic scheme of EV charging demand parameters.

Finally, the result of this methodology leads to the charging process model for each EV agent, the total EV charging demand and consequently, it allows the impact on power networks to be analyzed. The methodology proposed uses all sources from public data and it is applied using statistics from the city of Barcelona.

The EV charging demand model is defined as the electric demand from EVs during a certain time period, such as a day or week, to supply their batteries. EV charging demand depends on EV user driving needs and it is linked to EV characteristics and mobility of users.

The methodology proposed in this paper is the Agent-Based Modeling and Simulation (ABMS). The main strengths and applications of ABMS are listed as follows:

- Heterogeneous individual components: EV model and mobility pattern of each EV owner.
- Flexible systems: to manage the charging demand of each EV.
- Influence of location: to consider the effects of the charging point location in the power network.
- Representation of social interactions: different types of EV owners could have different influences on the total system.

For these reasons, this methodology has been used for obtaining EV mobility patterns with an heuristic approach [64]. Furthermore, this methodology enables to simulate complex systems; for instance, load demand in power systems [7] or virtual power plants to include different types of agents [65]. Thus, agent-based modeling has been selected for this research.

In this work, the EVs are a set of agents that has been defined as autonomous entities with their attributes and their processes are dynamic and time-dependent [66,67]. It allows defining each EV driver as an agent considering the usage of each vehicle. Each agent is simulated individually including possible interactions through the relationships between agents. Section 2 describes the characteristics of the agent-based model to obtain the charging demand from EVs and their impact on the distribution network.

2. EV CHARGING DEMAND MODEL

According to the Figure 1, the parameters needed to model the EV charging demand can be clustered in three groups: the EV agent (Section 2.1), mobility pattern (Section 2.2) and the charging process (Section 2.3). All these parameters permit the determination of all charging processes needed to reach each destination.

2.1. EV Agent

In the model developed, every EV agent represents an EV driver and its vehicle. The EV agent attributes are the EV model, the mobility needs, and the charging preferences. The EV agent behaviors are the trips taken (mobility), their corresponding energy consumption from their battery, the energy consumed from the electricity network to charge the battery, and the charging decision. For instance, when EV agents reach their destination, their charging process begin depending on the EV agent preferences and the energy price. The EV agent states with their corresponding variables are: waiting, driving, and charging.

Moreover, there are two other agents that influence on EV agents behavior: the Electricity Retailer Agent, who determines the electricity price for each instant, and the EV Aggregator Agent, who control the EV charges to reduce the electricity price. In the scenarios A, B and C, explained in the Section 3, there is no EV aggregator and the price is determined by the Electricity Retailer Agent. In contrast, in the scenario D, also explained in Section 3, the price is determined by the EV Aggregator Agent and the Electricity Retailer Agent does not influence on EV agents.

The main rule is that each EV agent, after each trip, takes the decision of charging in function of the battery state-of-charge, the electricity price and, in scenario D, the signal from the EV aggregator. Moreover, before changing the state of an EV agent from waiting to driving state, it is necessary that the battery has enough energy to reach the destination. The EV agents structure, their relationships with other agents and their environment are shown in Figure 2. Note that there are two environments related to the EV agents: spatial distribution and electricity network. Furthermore, the electricity market is the environment of Electricity Retailer Agent and EV Aggregator Agent.

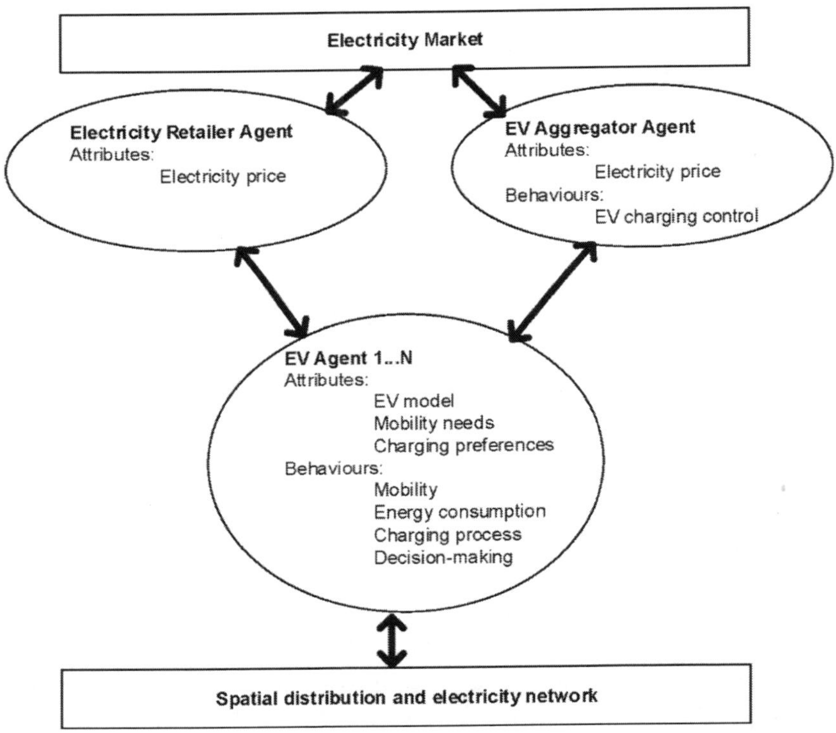

Figure 2. EV agent structure

When the simulation begins, the system computes the EV agent mobility needs and the battery state-of-charge variation.

The first step to define the EV agents is the definition of EV agent groups (C_i) and their variables. For each group, it is necessary to define the number of agents (N), spatial distribution of influence and charging preferences. And the EV model of each agent is defined with variables EC_i, Aut_i, Cap_i, Ps_i and $Type_i$. The place of residence, defined in R_i, is considered for each agent, and this depends on the power network scenario and is modeled as a constant probability, based on public data such as [68]. R_i is linked with the charging point in home usage.

The PDF of each EV model is based on [25] data and it just considers passenger vehicles and $Type_i$. This data was filtered for the case study in relation to EV model characteristics and technical data available from automakers. It is shown in Figure 3.

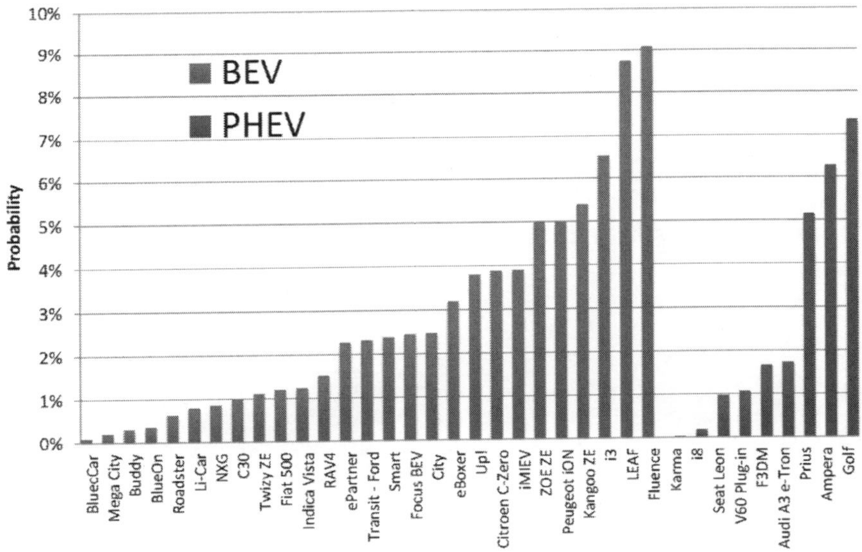

Figure 3. EV model probability distribution function EV_i. Based on [25] and adapted to Barcelona-Spain and automakers data.

In this model it is assumed that the PHEV drive is fully electric until the end of the energy stored in the battery, when they consume gasoline as hybrid electric vehicles. Other assumptions are exhibited in Section 2.4.

2.2. Mobility Pattern

Mobility variables are assigned to each EV agent in order to model its mobility behaviour. Different mobility patterns are based on open data sources. The variables considered to define a mobility pattern are defined as follows:

- **Trips per day** (S_i). The total trips are determined using a probabilistic variable which is generated through a Poisson distribution function, which is defined as [69] proposes with Poisson parameter (λ) of Equation (1).

$$P(k, \lambda) = (e^{-\lambda}\lambda^{k})/k! \tag{1}$$

  This parameter is based on the average statistic value. It should ensure at least two trips per day and is defined byEquation (2).

$$S_i = 2 + \lambda \tag{2}$$

  In the present study analyzed, $\bar{S}_i = 3.53$ trips/day are based on [70].

- **Distance** (L_i) and **Distance per trip** (l_{ij}). They are calculated using the exponential distribution function from public reports. Figure 4 shows cumulative exponential distribution functions of distance traveled per day from different countries and the relation between L_i and l_{ij} is shown the following equation:

$$L_i = \sum_{j=1}^{S_i} l_{ij} \tag{3}$$

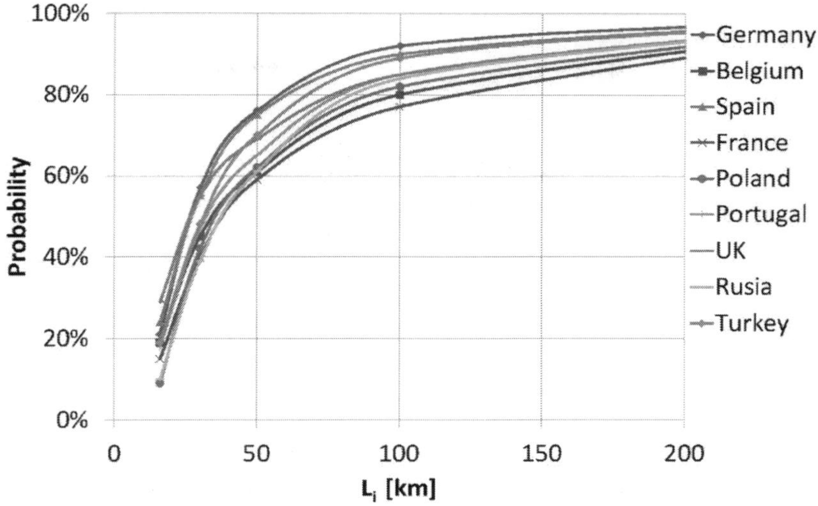

Figure 4. Probability distribution function of Distance L_i [71].

In the case study analyzed, $\bar{L_i} = 83\,\text{km/day}$ is based on [71]. If $l_{ij} > 10$ km, the trip j is considered as metropolitan considering Barcelona characteristics.

- **Destination** (D_{ij}). The model considers the reason of displacement to determine the destination. The reasons considered for the case study are based on the destination of each trip: for personal issues and for commuting. It is strongly linked to grid node, where the EV is connected in relation to social data and mobility pattern. The destination is modeled with a constant PDF according to the power network topology.

- **Day of the week** (d_i) and **Time distribution** (m_{ij}). These parameters allow knowing when an EV consumes energy as a function of the EV user's motivation to travel on a specific day. It is implemented in a PDF, as shown in Figure 5 and Table 1 as an example applied in the case study.

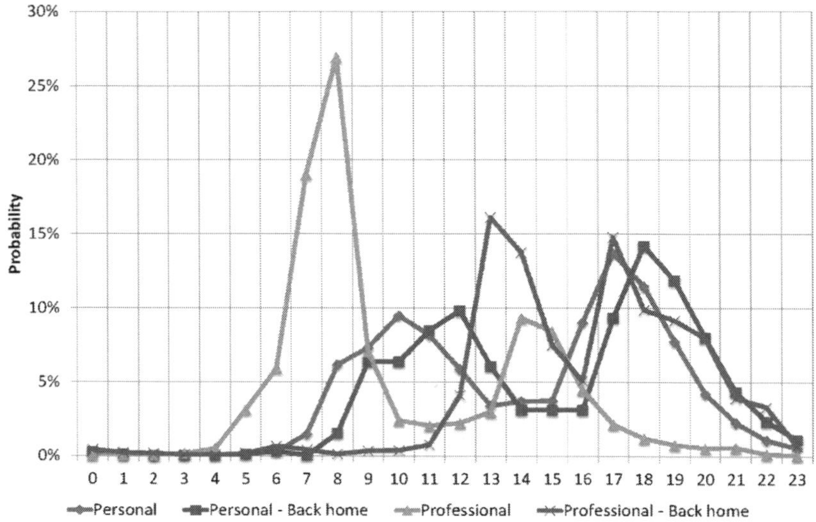

Figure 5. Probability distribution function of Time distribution m_{ij} [71].

Table 1. Time distribution considered in case study.

m_{ij}	**Description**
1	Personal
2	Personal—Back home
3	Professional
4	Professional—Back home

- **Velocity** (v_{ij}). According to mobility data, velocity is modelled as a constant value, depending if the trip is urban or metropolitan. The average velocity from [70] and $v^{urban} = 22.2$ km/h and $v^{metrop} = 59.3$ km/h are applied.

- **Initial/Final time** (t_0, t_1). The relation between them is the average velocity (v_{ij}) and distance (l_{ij}). Each pair of time variables is grouped in the matrix Y_i, which stores the mobility data of an EV agent.

$$Y_i = \begin{bmatrix} t_0^1 & t_1^1 \\ \vdots & \vdots \\ t_0^{S_i} & t_1^{S_i} \end{bmatrix}$$

- **Social variables.** Regarding the case study, it is necessary to take into account different variables such as Gross Domestic Product (GDP) and population density to determine the total number of agents (N) that could charge the EV at the same connection point. C_i definition was described in Section 2.1 and applied in Section 3.1.

2.3. Charging Process

The charging process considered is slow charging—AC single-phase, depending on EV model, battery capacity, SoC, Energy required to arrive to next destination and time between displacements.

All the EV models are supposed to have Li-ion batteries and the slow charging process corresponds to a typical charging curve with two periods: constant period I and descendant period II [72]. The power rate Ps_i considered for charging is 3.7 kW (230 V, 16 A) because it is commonly available in residential and commercial areas in Europe [73] and it is also used by Marra *et al.* [72]. The charging process depends on initial SoC and energy required (E_{req}) in the process. Figure 6 shows the charging process of a battery with Cap_i and E_{req} of 16.5 kWh.

In this model, it is assumed that period I requires 50% of time for a full charge and period II finishes when the power output reaches 8% of Ps_i.

- Total energy (Battery capacity) is: $Cap = E_I + E_{II}$.
- μ and k are the exponential function parameters used in Equation (6).
- Total process efficiency considered is 90% [2].

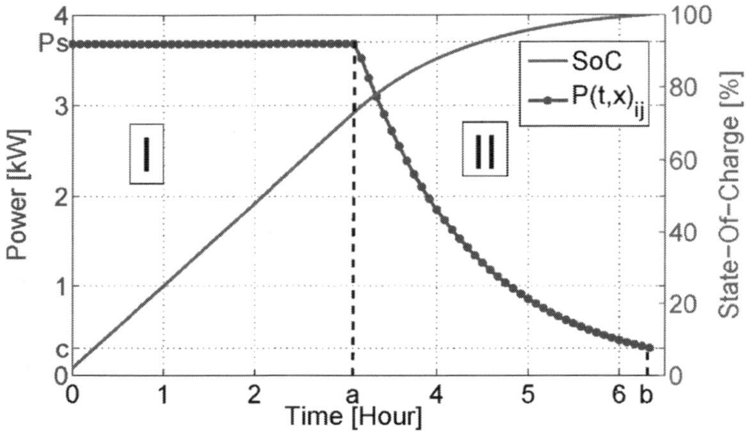

Figure 6. Slow charging profile—General scheme in relation to battery capacity. Based on [72].

The equations of EV charging process described before are:

- Period I is described by the following equations:

$$P_I(t) = Ps_i \tag{4}$$

$$E_I(t) = \int_0^a Ps_i dt \tag{5}$$

- Period II is described by the following equations:

$$P_{II}(t) = ke^{-\mu t} \tag{6}$$

$$E_{II}(t) = \int_a^b ke^{-\mu t} dt \tag{7}$$

where:

$$\mu = \frac{-ln(0.08)}{a} \tag{8}$$

$$k = \frac{Ps_i}{0.08} \tag{9}$$

$$c = 0.08Ps_i \tag{10}$$

The initial SoC depends on the EV agent consumption. In the first simulation, the battery starts fully charged.

2.4. Monte Carlo Simulation

Based on Figure 2, this paper proposes using the algorithm shown in Figure 7 to calculate the EV charging demand in a certain power network. This algorithm is based on Monte Carlo Methodology to include stochastic variables per agent and they are: R_i, S_i, L_i, l_{ij}, D_i, t_0, t_1 and EV_i. For this reason, it is necessary to define the number of iterations (T). Furthermore, to start the algorithm, it is necessary to define the number of agents (N) that charge the EV in the network analyzed. The time step used is 5 min.

The algorithm is used to define the EV agent group, the mobility variables and then the charging process for each EV agent.

3. CASE STUDY

The proposed EV charging demand model is applied in a case study with a 37-node IEEE test feeder adapted to a typical distribution network and mobility data of Barcelona (Spain) [70]. The modeling of the case study was implemented in MATLAB and the power flow is solved by means of the Newton-Raphson method.

Four charging scenarios (A-D) were defined to model EV agent behavior, which are described in the following sections. The results are the energy (Z_i) and charging demand from EVs ($P(t, x)$) and the voltage profile in the distribution network.

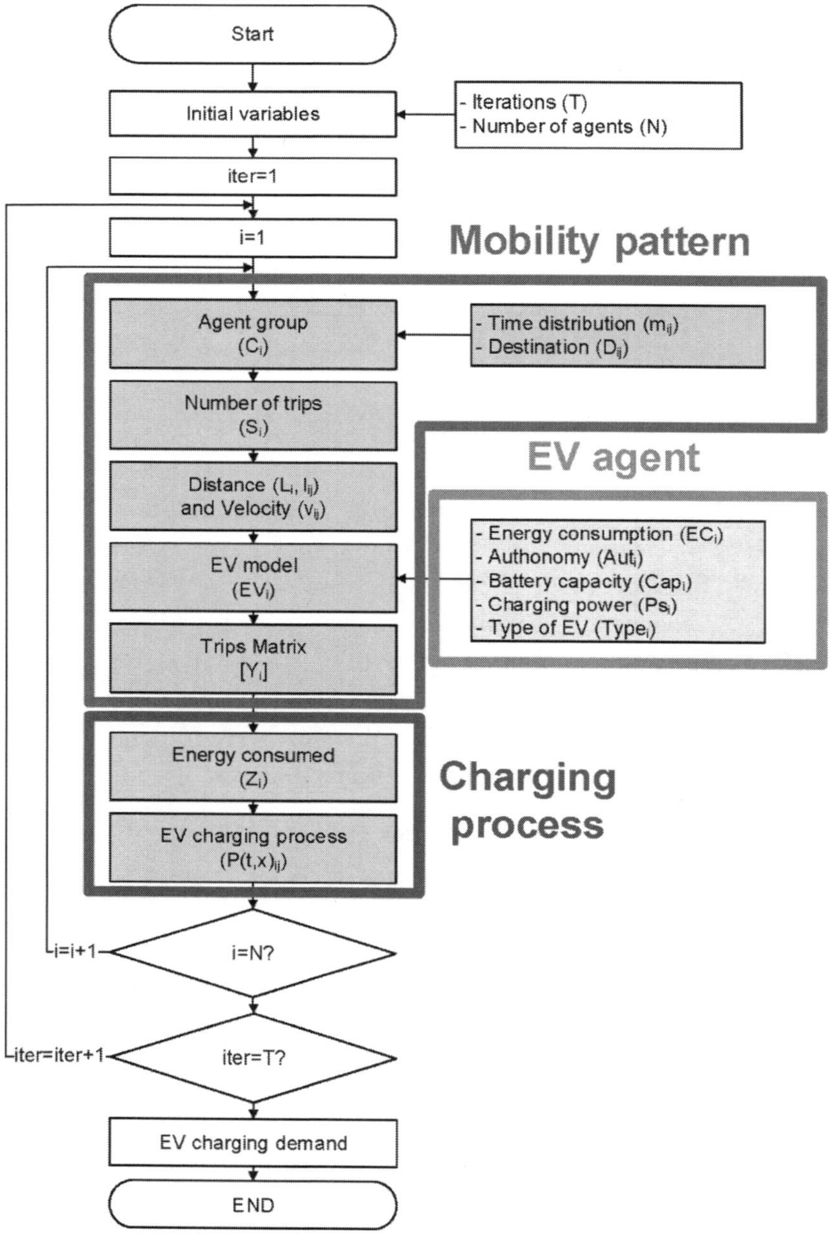

Figure 7. EV charging demand algorithm based on Monte Carlo.

3.1. Distribution Network

This case study is an adapted MV network 37-node IEEE test feeder, which is seen in Figure 8, and it applies Barcelona's mobility data. This network is adapted to a typical 25 kV MV network of Barcelona and the number of houses connected at the same MV/LV transformer [8]. In order to do that, it is necessary to consider social variables such as population density and technical regulation [74]. The maximum voltage drop permitted by the distribution system operator is 10% according to the EN 50160.

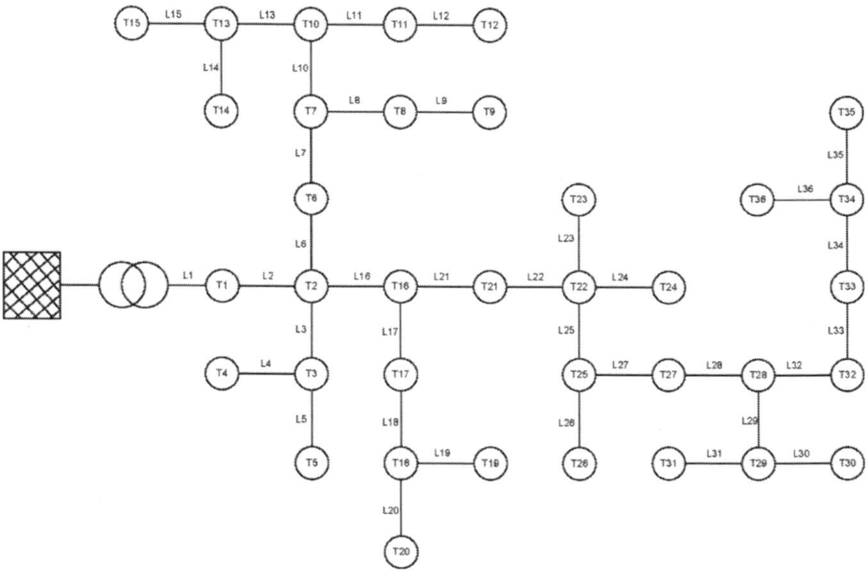

Figure 8. MV network - IEEE Test-feeder 37 node.

The total number of agents of group C_i is defined in relation to network topology and population density of different neighborhoods. According to social data from Barcelona and network branches, there are three zones: high, medium and low inhabitants per house and vehicles per inhabitant density. The farthest branch is linked with the high density zone. In this way, D_{ij} of group C_1 at the end of the day is the corresponding network node. In Barcelona, 38% of vehicles are driven each day and this percentage is used to determine active vehicles [70].

Table 2 shows calculations to get N of group C1.

Table 2. Number of agents C_1.

Zone	Nodes	Inhab./Hou.	Veh./Inhab.	Inhab.	Active Veh.
High	22–36	2.61	0.50	5016	950
Medium	3–5, 6–15	2.52	0.47	2541	448
Low	1, 2, 16–21	2.34	0.38	3288	471
				Total C1	**1870**

Load demand: Base load demand in this distribution network is based on system operator data [75] from national demand and it is adapted to network power capacity as 80% of HV/MV transformer power. Analyzing the consumption in Spain between 2007 and 2011, load demand used in the case study is from 17 December 2007, when the maximum energy demand reached 45,911 MWh between 18:00 and 19:00. This allows analyzing EV charging increase relative to this base load.

The load presented in Figure 9a is the base case, without EVs, of the distribution system analyzed. The peak demand is 10,640 kW and it occurs at 18:30. The load demand of the distribution system increases during the morning (8–10 o'clock), decreases during lunch time (13–16 o'clock) and increases during the evening (19–21 o'clock), when people come back home. The peak period is 79% higher than the valley period and the energy consumed during the course of a single day is 207.36 MWh. The voltage in the worst node is shown in Figure 9b; the minimum voltage is 0.9707 p.u. at 18:30 and the maximum is 0.9839 p.u. at 4:45. The voltage follows a similar behavior to the load demand. The lower limit of the voltage magnitude permitted by EN 50160 is 0.90 p.u.

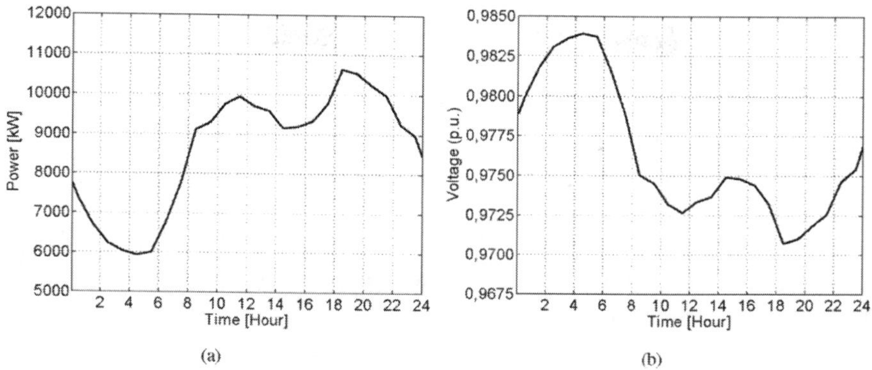

Figure 9. Residential and commercial demand without EVs. (**a**) Load demand; (**b**) Voltage drop.

3.2. Agent Profile

Six agent groups (C1–C6) were defined to consider mobility and residence. Mobility is divided between personal and professional reasons. According to the usual place where the EV is connected at the end of the day, three different areas of residence were defined: local, urban and metropolitan. Local area refers to the distribution network analyzed, urban refers to the city, and metropolitan is outside the city. Urban and metropolitan agents can plug in between displacements. On the other hand, local agents can charge at any time. Table 3 shows the main characteristics of each group. N is the number of EVs of each agent that charge their batteries in the case study network.

Table 3. EV charging social characteristics in function of group.

C_i	m_{ij}	Active Veh.	N	Area	Preferences
C1	1 & 2	1870	561	Local	At-the-end
C2	1 & 2	449	135	Urban	Between disp.
C3	1 & 2	273	82	Metropolitan	Between disp.
C4	3 & 4	41	12	Local	At-the-end
C5	3 & 4	41	12	Urban	Between disp.
C6	3 & 4	10	3	Metropolitan	Between disp.
Total		2684	805		

Each group has specific energy requirements for charging (E_{req}). Preferences are related to when to charge and they are described above relative to agent group definition. Regarding the E_{req} for each feasible charge between displacements, it is defined as the energy required to reach the next destination (D_{ij}) and distance (l_{ij}).

Mobility variables from Barcelona data [70] are implemented in the case study. S_i depends on agent group, d_i is the average weekday and L_i is according to [71].

3.3. Charging Scenarios

According to agent preferences, E_{req} and electricity market assumptions, four scenarios of EV charging demand are described, shown in Table 4.

Table 4. Table of charging scenarios.

Charging Scenario	Description	Range Anxiety
A—Intensive charge	As soon as possible	High
B—Plug-and-Play	Just at home	Medium
C—Tariff controlled	Off-peak tariff	Medium
D—Smart charging	With Aggregator	Low

Scenarios A and B consider constant electricity price for the whole day. In scenario A, EVs charge at the end of each trip due to the high range anxiety of EV agents. In scenario B, the EV agents have lower range anxiety and they charge the vehicle at home, when SoC is lower than 20% or lower than E_{req}. In scenario C it is considered that the EV agents have a Time-of-Use (TOU) tariff, special for EVs [76]. The cheapest period of this tariff begins at 1:00 am, based on the Spanish regulation [77], and then the EVs initiate the charge. The TOU tariff is an indirect control strategy to manage the EV charges. Scenario D considers one aggregator who manages all EV charges to consume the minimum power at the HV/MV transformer. This is based on an aggregator dedicated to reducing the impact in the transmission system, according to the Spanish regulation [77]. This scenario shows a direct control strategy to manage the EV charges and the aggregator offers lower electricity prices for EV agents.

3.4. Results

The following discussion presents the results of the four scenarios simulated. The analysis is focused on the EV demand, total demand and the voltage drop in the worst node. Due to the probabilistic design of the model, the results are variable and the plots show the variation between the maximum and minimum energy consumption. Furthermore, the plots also show the average consumption as the most probable value.

All scenarios are simulated considering that 30% of active vehicles are electric (N), based on maximum scenarios in [2,22,78]. EV_i PDF is based on [25]. What is also considered is that the EV agents with the value L_i greater than 100 km are only PHEV ($Type_i$).

The impact on power system is analyzed through voltage drop located in the farthest node, which is the 35. Figure 10shows the minimum voltage per node during the whole day and the maximum voltage drop is located in node 35.

Iterations (*iter*). The standard deviations (std. dev.) of power demand are evaluated to determine the number of iterations (T) to obtain valid results. To do that, a simulation with 1200 iterations in scenario A for C1 group and with 30% of EVs was carried out.

Figure 11 shows the std. dev. around hour 21 and it varies during the first 100 iterations significantly; it is nearly stable from iteration 200 and is constant from iteration 600. The ideal should be to do 600 iterations for all the cases, but the computing time to do it is very high and the volume of results to be stored requires a huge amount of memory. For these reasons, it is not possible to simulate 600 iterations for all the scenarios and the number of iterations has to be lower. The std. dev. varies around 10 kW from iteration 100 and from iteration 200, the results are more stable than previously. According to this, the number of iterations applied in the case study is 200. Other instances and scenarios are also checked and they comply with the std. dev. analysis. The consumption variation is also checked and it behaves similarly to the std. dev.

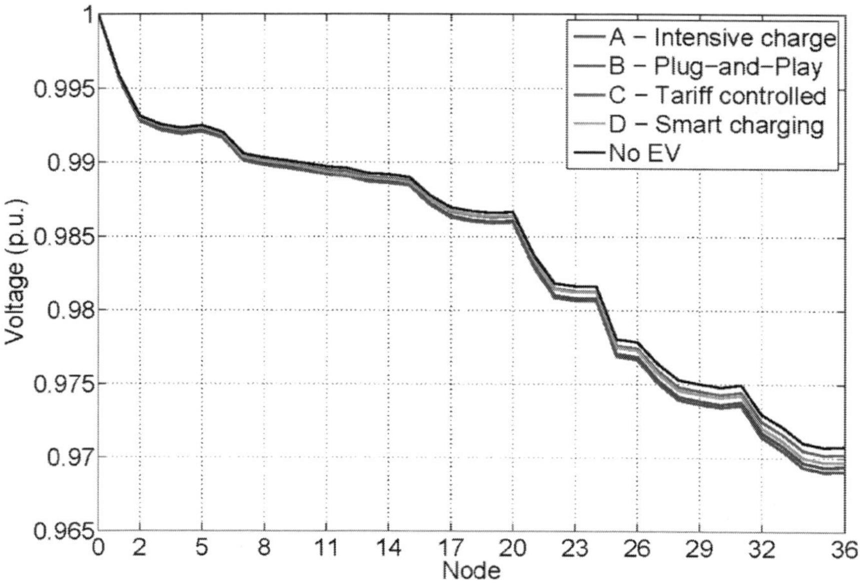

Figure 10. Voltage per node.

Figure 10. Voltage per node.

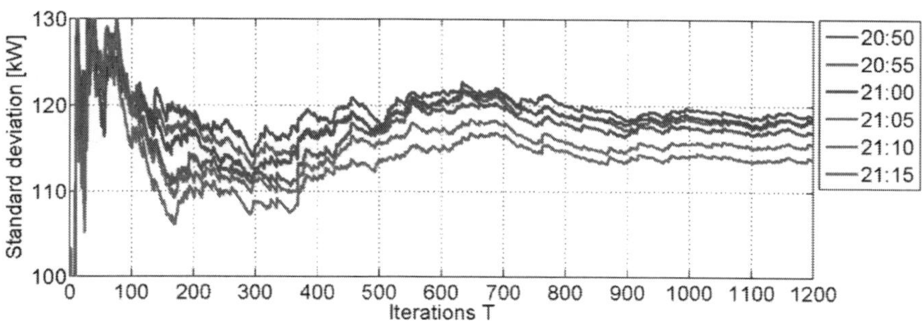

Figure 11. Standard deviation variation in function of iterations T.

3.4.1. A—Intensive Charge

EV charging demand: As is shown in Figure 12a, the EV charging demand presents two peaks with more consumption around 10:00 and 19:00. Both peaks are related to Barcelona's mobility pattern illustrated in Figure 5, which shows the same peaks: the peak during the morning is caused by professional mobility and the peak during the evening is caused by professional and personal back home reasons. The EV charging demand variability, the difference between the minimum and the maximum case, is significant in this scenario, and it can reach the 50% of the EV consumption as it occurs at 20:00.

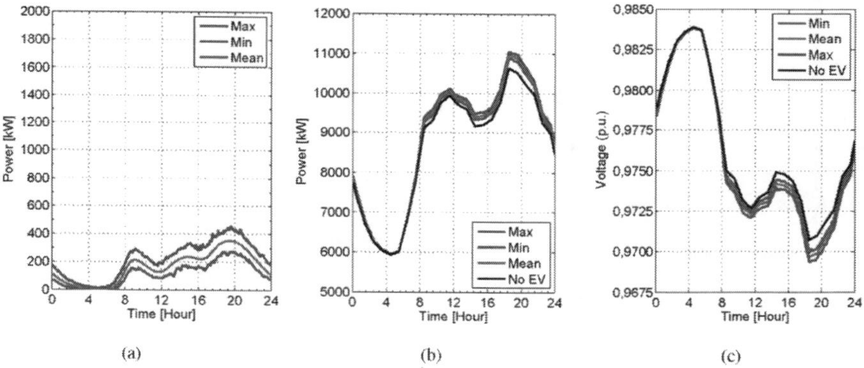

Figure 12. A—Intensive charge. (**a**) EV charging demand; (**b**) Total demand; (**c**) Voltage drop.

The EV peak demand is near to 500 kW and the total peak demand is 11.04 MW, 3.75% higher than in the base case without EVs, as Figure 12b demonstrates. Furthermore, the peak during the morning is coupled with the residential and commercial demand. This is reflected in Figure 12b, where the active power increase is steeper from 6 to 12 hours due to the EV charging demand.

Impact on power system: Figure 12c shows that the minimum voltage in node 35 is 0.9694 p.u. and it is 0.13% lower than in the No EV case, which is higher than the lower limit of the standard of 0.9 p.u.

3.4.2. B—Plug-and-Play

EV charging demand: In this scenario, the EV agents prefer to charge at home, according to the back home time distributions (m_{ij}). As shown in Figure 13a, the first peak demand is lower than in scenario A because the agents do not charge at work. Moreover, the second peak demand is higher than before because the agents have not charged at work and the energy required by them is higher than in scenario A. In this scenario, the EV charging demand variability is also significant and it can reach the 33% of the EV consumption, as it occurs at 20:00.

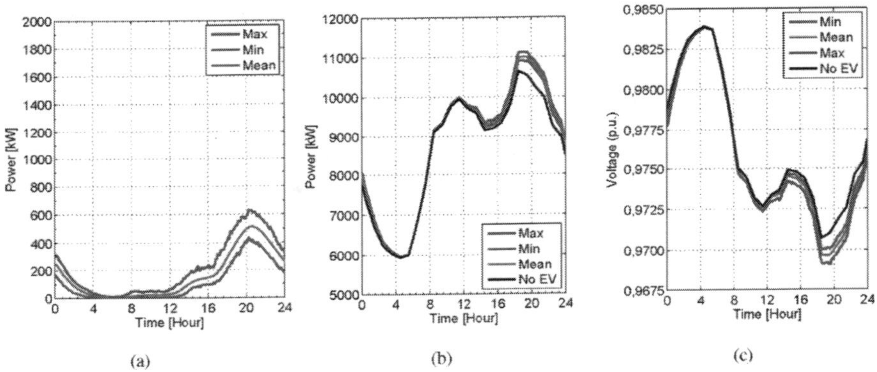

Figure 13. B—Plug-and-Play. (**a**) EV charging demand; (**b**) Total demand; (**c**) Voltage drop.

As Figure 13b shows, this effect causes that the peak during the morning in the total demand is lower than the previous case. And the peak during the evening is higher due to the energy required and the maximum power consumed is 11,12 MW at 18:35 and the relative increase from the case without EV is 4.51%. Moreover, the power consumption during the night is higher than in case A, because the SoC of EV agents when they arrive at home is lower than previously.

Impact on power system: Figure 13c shows that the combination of the peak from the residential demand with the EV demand causes a higher voltage drop than scenario A, due to the different behaviors of the EV agents. The minimum voltage reached during the peak demand is 0.9691 p.u., 0.16% lower than the case without EV, and higher than the lower limit of 0.90 p.u.

3.4.3. C—Tariff Controlled

EV charging demand: In this case, the TOU tariff causes that the EV agents begin to charge at 1:00, when the energy is cheaper. Therefore, the EV charging demand presents a peak of 1.86 MW at this moment due to the simultaneous EV charges, as seen in Figure 14a. What is more, the control reduces the EV charging demand variability.

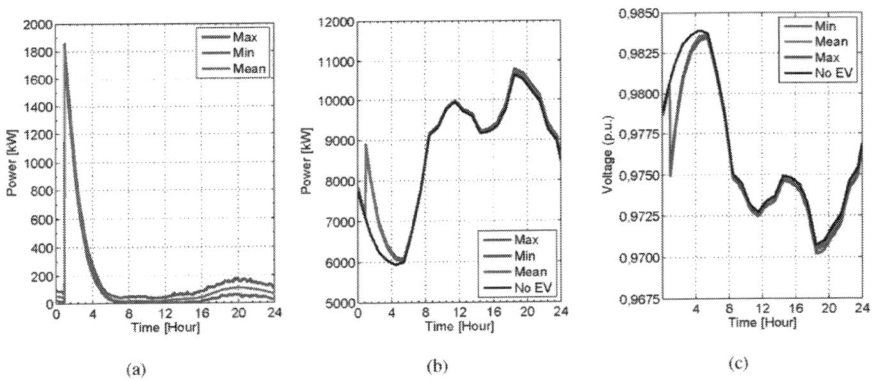

Figure 14. C—Tariff controlled. (**a**) EV charging demand; (**b**) Total demand; (**c**) Voltage drop.

The consumption during the rest of the day is related to the energy required (E_{req}) to reach the next destination (D_{ij}) and the low SoC of each EV agent. The maximum power consumed is 10.8 MW at 18:30, which means an increase of 1.5% from the original case.

Figure 14b shows that this EV peak happens during the off-peak period and the total demand increase is not significant. Despite this, the power generation gradient could be a problem, which should be analysed from the point of view of the power generation and from the system stability point of view.

Impact on power system: The minimum voltage, shown in Figure 14c, is similar to the original case without EVs. The minimum voltage reached is 0.9702 p.u., 0.05% lower than without EVs, and higher than 0.90 p.u. The voltage variation at 1:00 could be a problem, which could be analyzed in a transient analysis.

3.4.4. D—Smart Charging

EV charging demand: Figure 15a shows the EV charging demand controlled by the aggregator which controls domestic EV charges. The EV charging demand is shifted to the valley period to reduce the consumption through the HV/MV transformer and to minimize the impact on the transmission system. According to this, the EV charges occur between 2 and 8 o'clock and the variability, the difference between the minimum and the maximum case, is very small.

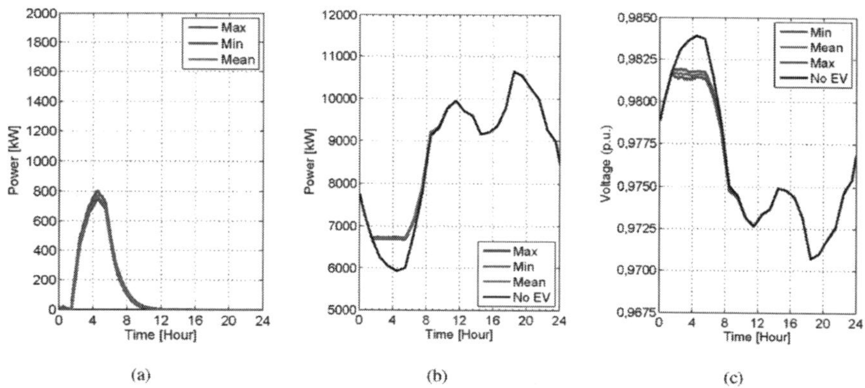

Figure 15. D—Smart charging. (**a**) EV charging demand; (**b**) Total demand; (**c**) Voltage drop.

Figure 15b shows that the total demand increases during the valley periods and the power consumption is constant at 6.6 MW. During the rest of the day, sporadic charges could occur, but the mean curve is near to the case without EVs.

Impact on power system: The minimum voltage is not increased by the EV charges, as is exhibited in Figure 15c. The voltage during the valley period is lower than in the original case according to the total demand, but this voltage is higher than during the peak hours, and the difference between the minimum voltages is 0.02%, and the minimum value of 0.90 p.u. is not reached.

The summary of all the scenarios is presented in Table 5. Voltage value is the minimum and it means the maximum voltage drop.

Table 5. Maximum results.

Scenario	EV Demand (Max) [kW]	Peak Time	Total Demand (Max) [kW]	Variation	Peak Time	Voltage (Min) [p.u.]	Variation
No EV		18:30	10640		18:30	0.9707	
A—Intensive charge	457	18:30	11040	3.76%	18:30	0.9694	-0.13%
B—Plug-and-Play	628	18:35	11120	4.51%	18:35	0.9691	-0.16%
C—Tariff controlled	1857	01:00	10800	1.50%	18.30	0.9702	-0.05%
D—Smart charging	799	04:30	10720	0.75%	18:35	0.9705	-0.02%

Box plots Figure 16a–d show total consumption in each node and this is compared to MV/LV transformer capacity. The results show that the nodes with less capacity could reach the nominal value in some cases, but the average value is under nominal power. In the case of scenario D, total demand never exceeds the nominal capacity of transformers, which means that there is enough capacity to supply the EVs.

4. CONCLUSIONS

The probabilistic agent-based model (ABM) obtained in this paper allows the EV charging demand to be determined, taking into account different variables of EV characteristics such as battery capacity and energy consumption of each trip, economic and social attributes, mobility needs, and charging strategies of each agent. The model developed takes into account the interaction of these variables, allowing the obtainment of better accuracy in the results.

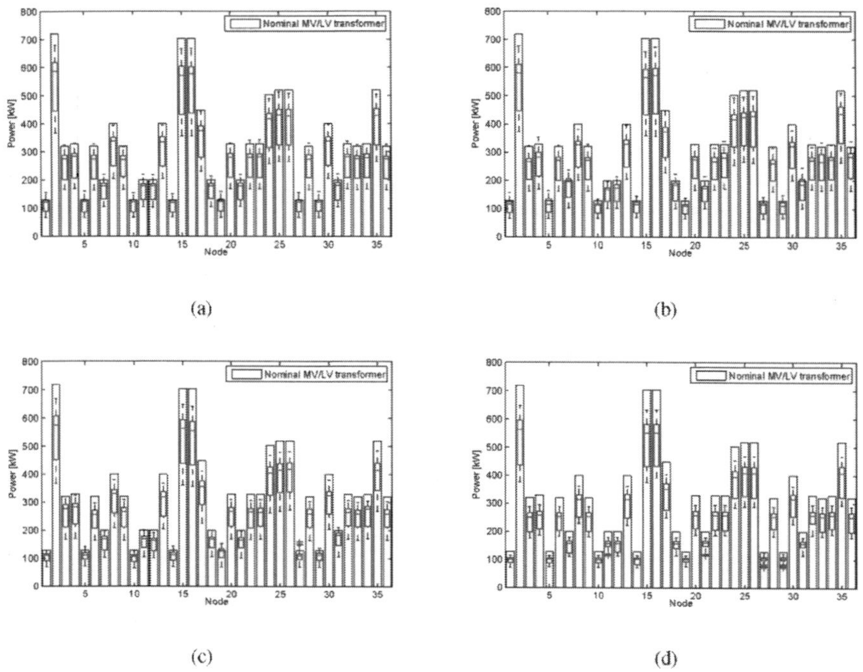

Figure 16. Total demand in each MV/LV transformer. (**a**) A—Intensive charge; (**b**) B—Plug-and-Play; (**c**) C—Tariff controlled; (**d**) D—Smart charging.

The probabilistic approach is useful to include the uncertainties related to the real behavior of EV users, like the time distribution and energy consumed on each trip. Therefore, the model permits the determination of the impact provoked on the grid by these uncertainties.

Moreover, the model proposed is a benchmark to compare case studies, such as different cities or areas in the same city. With this model, the weak regions of the grid or the areas with high EV density can be detected.

The case study presented shows that the uncertainties cause variability in the EV charging demand in scenarios without control on the EVs, as it is shown in scenarios A and B. In contrast, the consumption variability in scenarios with indirect and direct control on the EV charges, like scenarios C and D, respectively, is small.

The distribution feeder analyzed in the presented case study does not have a significant impact on the smart charging strategy (D) during the off-peak period and all EV agents can charge their EV. In contrast, some MV/LV transformers could exceed their nominal power in the scenarios without control. The voltage in all the scenarios is higher than the limit of 0.90 p.u. according to the EN 50160.

In further work, it could be very interesting to analyze the dynamic behavior of the system in case C during the connection of all EV at 1:00. Furthermore, the model permits analyzing the impact on distribution networks, but it can be applied for transmission and low voltage grids, too. Finally, this model could be applied and compared with a real distribution network with EVs to verify the accuracy of the model.

ACKNOWLEDGMENTS

This research was co-financed by the Spanish Ministry of Economy and Competitiveness (MINECO) through the research project ENE2012-36337 "Micro redes con almacenamiento para la integración óptima de energías renovables y vehiculos electricos". This project has received funding from the European Union's Horizon 2020 research and innovation programme under grant agreement No 646476.

AUTHOR CONTRIBUTIONS

Pol Olivella-Rosell, Roberto Villafafila-Robles and Andreas Sumper conceived the model; Pol Olivella-Rosell performed the simulations; Pol Olivella-Rosell, Roberto Villafafila-Robles, Andreas Sumper and Joan Bergas-Jané analysed the data; Pol Olivella-Rosell wrote the paper.

CONFLICTS OF INTEREST

The authors declare no conflict of interest.

REFERENCES

1 Bullis, K. Could Electric Cars Threaten the Grid? In MIT Technology Review; Massachusetts Institute of Technology (MIT): Cambridge, MA, USA, 2013.

2 Clement-Nyns, K.; Haesen, E.; Driesen, J. The Impact of Charging Plug-In Hybrid Electric Vehicles on a Residential Distribution Grid. IEEE Trans. Power Syst. 2010, 25, 371–380.

3 Huang, S.; Infield, D. The impact of domestic Plug-in Hybrid Electric Vehicles on power distribution system loads, Proceedings of the 2010 International Conference on Power System Technology (POWERCON), Hangzhou, China, 24–28 October 2010.

4 Villafafila-Robles, R.; Girbau-Llistuella, F.; Olivella-Rosell, P.; Sudria-Andreu, A.; Bergas-Jane, J. Assessment of the impact of charging infrastructure for electric vehicles on distribution networks, Proceedings of the 15th European Conference on Power Electronics and Applications, Lille, France, 2–6 September 2013.

5 Watts, R.A.; Letendre, S. The Effects of Plug-in Hybrid Electric Vehicles on the Vermont Electric Transmission System, Proceedings of the Transportation Research Board 88th Annual Meeting, Washington, DC, USA, 11–15 January 2009.

6 Soãres, J.; Canizes, B.; Lobo, C.; Vale, Z.; Morais, H. Electric Vehicle Scenario Simulator Tool for Smart Grid Operators. Energies 2012, 5, 1881–1899.

7 Acha, S.; van Dam, K.H.; Shah, N. Modelling Spatial and Temporal Agent Travel Patterns for Optimal Charging of Electric Vehicles in Low Carbon Networks, Proceedings of the IEEE Power and Energy Society General Meeting, San Diego, CA, USA, 22–26 July 2012.

8 Valsera-Naranjo, E.; Sumper, A.; Villafafila-Robles, R.; Martinez-Vicente, D. Probabilistic Method to Assess the Impact of Charging of Electric Vehicles on Distribution Grids. Energies 2012, 5, 1503–1531.

9 Soares, F.J.; Lopes, J.A.P.; Almeida, P.M.R.; Moreira, C.L.; Seca, L. A stochastic model to simulate electric vehicles motion and quantify the energy required from the grid, Proceedings of the Power Systems Computation Conference (PSCC), Stockholm, Sweden, 22–26 August 2011.

10 Amjad, S.; Neelakrishnan, S.; Rudramoorthy, R. Review of design considerations and technological challenges for successful development and deployment of plug-in hybrid electric vehicles. Renew. Sustain. Energy Rev. 2010, 14, 1104–1110.

11 Peng, M.; Liu, L.; Jiang, C. A review on the economic dispatch and risk management of the large-scale plug-in electric vehicles PHEVs-penetrated power systems. Renew. Sustain. Energy Rev. 2012, 16, 1508–1515.

12 Gao, S.; Chau, K.T.; Wu, D.; Chan, C.C. Modeling and coordinated control for integrating electric vehicles into the power grid, Proceedings of the 2011 International Conference on Electrical Machines and Systems (ICEMS), Beijing, China, 20–23 August 2011.

13 Sortomme, E.; Hindi, M.M.; MacPherson, S.D.J.; Venkata, S.S. Coordinated Charging of Plug-In Hybrid Electric Vehicles to Minimize Distribution System Losses. IEEE Trans. Smart Grid. 2011, 2, 198–205.

14 Wang, J.; Liu, C.; Ton, D.; Zhou, Y.; Kim, J.; Vyas, A. Impact of plug-in hybrid electric vehicles on power systems with demand response and wind power. Energy Policy 2011, 39, 4016–4021.

15 Waraich, R.A.; Galus, M.D.; Dobler, C.; Balmer, M.; Andersson, G.; Axhausen, K.W. Plug-in hybrid electric vehicles and smart grids: Investigations based on a microsimulation. Transp. Res. Part C Emerg. Technol. 2013, 28, 74–86.

16 Rahman, S.; Shrestha, G.B. An investigation into the impact of electric vehicle load on the electric utility distribution system. IEEE Trans. Power Deliv. 1993, 8, 591–597.

17 Valsera-Naranjo, E.; Martinez-Vicente, D.; Sumper, A.; Villafafila-Robles, R.; Sudria-Andreu, A. Deterministic and probabilistic assessment of the impact of the electrical vehicles on the power grid, Proceedings of the 2011 IEEE Power and Energy Society General Meeting, San Diego, CA, USA, 24–29 July 2011.

18 Mateo, A.R. Evaluación del impacto de los vehículos eléctricos en las redes de distribución. Master Thesis, Universidad Pontificia de Comillas—Escuela técnica superior de ingeniería ICAI, Madrid, Spain, 2010.

19 Druitt, J.; Früh, W.G. Simulation of demand management and grid balancing with electric vehicles. J. Power Sources.2012, 216, 104–116.

20 Loisel, R.; Pasaoglu, G.; Thiel, C. Large-scale deployment of electric vehicles in Germany by 2030: An analysis of grid-to-vehicle and vehicle-to-grid concepts. Energy Policy 2014, 65, 432–443.

21 Lyon, T.P.; Michelin, M.; Jongejan, A.; Leahy, T. Is smart charging policy for electric vehicles worthwhile? Energy Policy 2012, 41, 259–268.

22 Maitra, A.; Taylor, J.; Brooks, D.; Alexander, M.; Duvall, M. Integrating plug-in-electric vehicles with the distribution system, Proceedings of the 20th International Conference and Exhibition on Electricity Distribution—Part 1, Prague, Czech Republic, 8–11 June 2009.

23 Lopes, J.A.P.; Soares, F.J.; Almeida, P.M.R. Identifying management procedures to deal with connection of Electric Vehicles in the grid, Proceedings of the 2009 IEEE Bucharest PowerTech, Bucharest, Romania, 28 June–2 July 2009.

24 Pang, C.; Dutta, P.; Kezunovic, M. BEVs-PHEVs as Dispersed Energy Storage for V2B Uses in the Smart Grid. IEEE Trans. Smart Grid. 2012, 3, 473–482.

25 Frost & Sullivan, Executive Analysis of Global Electric Vehicle Forecast; Technical Report; Frost & Sullivan: Mountain View, CA, USA, 2012.

26 Tomić, J.; Kempton, W. Using fleets of electric-drive vehicles for grid support. J. Power Sources. 2007, 168, 459–468.

27 Metz, M.; Doetsch, C. Electric vehicles as flexible loads—A simulation approach using empirical mobility data. Energy2012, 48, 369–374.

28 López, M.; Martín, S.; Aguado, J.; de la Torre, S. V2G strategies for congestion management in microgrids with high penetration of electric vehicles. Electr. Power Syst. Res. 2013, 104, 28–34.

29 Electric Vehicle Conductive Charging System—Part 1: General Requirements; IEC Standard 61851-1; International Electrotechnical Commission: Geneva, Switzerland, 2010.

30 Vehicle Conductive Charge Coupler; SAE Standard J1772; Society of Automotive Engineers: California Air Resources Board, CA, USA, 2009.

31 Geth, F.; Debreucker, S.; Clement, K.; Driesen, J. Charging Power Analysis for a Belgian Plug-in Hybrid Electric Vehicle Fleet, Proceedings of the 5th IEEE Young Researchers Symposium, Leuven, Belgium, 29–30 March 2010.

32 Zhang, L.; Brown, T.; Samuelsen, G.S. Fuel reduction and electricity consumption impact of different charging scenarios for plug-in hybrid electric vehicles. J. Power Sources. 2011, 196, 6559–6566.

33 Grenier, A.; Page, S. The impact of electrified transport on local grid infrastructure: A comparison between electric cars and light rail. Energy Policy 2012, 49, 355–364.

34 Collins, M.M.; Mader, G.H. The timing of EV recharging and its effect on utilities. IEEE Trans. Veh. Technol. 1983, 32, 90–97.

35 Clement-Nyns, K.; Haesen, E.; Driesen, J. The impact of vehicle-to-grid on the distribution grid. Electr. Power Syst. Res.2011, 81, 185–192.

36 Guo, Q.; Wang, Y.; Sun, H.; Li, Z.; Xin, S.; Zhang, B. Factor Analysis of the Aggregated Electric Vehicle Load Based on Data Mining. Energies 2012, 5, 2053–2070.

37 Qian, K.; Zhou, C.; Allan, M.; Yuan, Y. Modeling of Load Demand Due to EV Battery Charging in Distribution Systems. IEEE Trans. Power Syst. 2011, 26, 802–810.

38 Gao, S.; Chau, K.; Chan, C.C.; Liu, C.; Wu, D. Optimal Control Framework and Scheme for Integrating Plug-in Hybrid Electric Vehicles into Grid. J. Asia Electr. Veh. 2011, 9, 1473–1481.

39 Lojowska, A.; Kurowicka, D.; Papaefthymiou, G.; van der Sluis, L. From transportation patterns to power demand: Stochastic modeling of uncontrolled domestic charging of electric vehicles, Proceedings of the 2011 IEEE Power and Energy Society General Meeting, San Diego, CA, USA, 24–29 July 2011.

40 Multin, M.; Allerding, F.; Schmeck, H. Integration of electric vehicles in smart homes—An ICT-based solution for V2G scenarios, Proceedings of the 2012 IEEE PES Innovative Smart Grid Technologies (ISGT), Washington, DC, USA, 16–20 January 2012.

41 Garcia-Valle, R.; Vlachogiannis, J.G. Letter to the Editor: Electric Vehicle Demand Model for Load Flow Studies. Electr. Power Compon. Syst. 2009, 37, 577–582.

42 Keirstead, J.; Jennings, M.; Sivakumar, A. A review of urban energy system models: Approaches, challenges and opportunities. Renew. Sustain. Energy Rev. 2012, 16, 3847–3866.

43 Kelly, J.C.; MacDonald, J.S.; Keoleian, G.A. Time-dependent plug-in hybrid electric vehicle charging based on national driving patterns and demographics. Appl. Energy 2012, 94, 395–405.

44 Stephens, T. An Agent-Based Model of Energy Demand and Emissions From Plug-In Hybrid Electric Vehicle Use; Technical Report; University of Michigan: Ann Arbor, MI, USA, 2010.

45 Weiller, C. Plug-in hybrid electric vehicle impacts on hourly electricity demand in the United States. Energy Policy2011, 39, 3766–3778.

46 Huang, S.; Infield, D. The potential of domestic electric vehicles to contribute to Power System Operation through vehicle to grid technology, Proceedings of the 44th International Universities Power Engineering Conference (UPEC), Glasgow, UK, 1–4 September 2009.

47 Huang, S.; Infield, D. Demand side management for domestic plug-in electric vehicles in power distribution system operator, Proceedings of the 21st International Conference and Exhibitionon Electricity Distribution, Frankfurt, Geramny, 6–9 July 2011.

48 Schroeder, A.; Traber, T. The economics of fast charging infrastructure for electric vehicles. Energy Policy 2012, 43, 136–144.

49 Juul, N.; Meibom, P. Road transport and power system scenarios for Northern Europe in 2030. Appl. Energy 2012, 92, 573–582.

50 Mora, R.; Oyarzabal, J.; Cruz-Zambrano, M.; Gonzalez, A.; Corera, J. E-car and economic impact: Enhancing the smart grids, Proceedings of the 2012 Workshop Integration of Renewables into the Distribution Grid, Lisbon, Portugal, 29–30 May 2012.

51 Waraich, R.; Galus, M.D.; Dobler, C.; Balmer, M.; Andersson, G.; Axhausena, K.W. Plug-in Hybrid Electric Vehicles and Smart Grid: Investigations Based on a Micro-Simulation, Proceedings of the 12th International Conference on Travel Behaviour Research (IATBR), Jaipur, India, 13–18 December 2009.

52 Galus, M.D.; Andersson, G. Demand Management of Grid Connected Plug-In Hybrid Electric Vehicles (PHEV), Proceedings of the Energy 2030 Conference, Atlanta, GA, USA, 17–18 November 2008.

53 Venkatesan, N.; Solanki, J.; Solanki, S.K. Residential Demand Response model and impact on voltage profile and losses of an electric distribution network. Appl. Energy 2012, 96, 84–91.

54 Galus, M.; Waraich, R.; Balmer, M.; Andersson, G.; Axhausen, K. A framework for investigating the impact of PHEVS, Proceedings of the International Advanced Mobility Forum (IAMF), ZurichETH,

Swiss Federal Institute of Technology, IVT Institute for Transport Planning and Systems, Geneva, Switzerland, 10–12 March 2009.

55 Galus, M.; Andersson, G. Integration of Plug-In Hybrid Electric Vehicles into energy networks, Proceedings of the 2009 IEEE PowerTech Bucharest, Bucharest, Romania, 28 June–2 July 2009.

56 Balmer, M. Travel demand modeling for multi-agent traffic simulations: Algorithms and systems. Ph.D. Thesis, ETH Zurich, Zurich, Switzerland, 2007.

57 Hedegaard, K.; Ravn, H.; Juul, N.; Meibom, P. Effects of electric vehicles on power systems in Northern Europe.Energy 2012, 48, 356–368.

58 Mohsenian-Rad, A.; Wong, V.W.S.; Jatskevich, J.; Schober, R.; Leon-Garcia, A. Autonomous Demand-Side Management Based on Game-Theoretic Energy Consumption Scheduling for the Future Smart Grid. IEEE Trans. Smart Grid. 2010, 1, 320–331.

59 Smith, R.; Shahidinejad, S.; Blair, D.; Bibeau, E.L. Characterization of urban commuter driving profiles to optimize battery size in light-duty plug-in electric vehicles. Transp. Res. Part D Transp. Environ. 2011, 16, 218–224.

60 Kleiwegt, E.; Lukszo, Z. Grid Impact Analysis of Electric Mobility on a Local Electricity Grid, Proceedings of the 9th IEEE International Conference on Networking, Sensing and Control (ICNSC), Beijing, China, 11–14 April 2012.

61 Kempton, W.; Tomic, J. Vehicle-to-grid power fundamentals: Calculating capacity and net revenue. J. Power Sources.2005, 144, 268–279.

62 Zakariazadeh, A.; Jadid, S.; Siano, P. Multi-objective scheduling of electric vehicles in smart distribution system.Energy Convers. Manag. 2014, 79, 43–53.

63 Dallinger, D.; Wietschel, M. Grid integration of intermittent renewable energy sources using price-responsive plug-in electric vehicles. Renew. Sustain. Energy Rev. 2012, 16, 3370–3382.

64 ElBanhawy, E.; Dalton, R.; Thompson, E.; Kotter, R. A heuristic approach for investigating the integration of electric mobility charging infrastructure in metropolitan areas: An agent-based modeling simulation, Proceedings of the 2012 2nd International Symposium on Environment Friendly Energies and Applications (EFEA), Newcastle upon Tyne, UK, 25–27 June 2012.

65 Vale, Z.; Pinto, T.; Morais, H.; Praça, I.; Faria, P. VPP's multi-level negotiation in smart grids and competitive electricity markets, Proceedings of the IEEE Power and Energy Society General Meeting, San Diego, CA, USA, 24–29 July 2011.

66 Macal, C.; North, M. Tutorial on agent-based modelling and simulation. J. Simul. 2010, 4, 151–162.

67 Bonabeau, E. Agent-based modeling: Methods and techniques for simulating human systems. Proc. Natl. Acad. Sci. USA 2002, 99, 7280–7287.

68 Barcelona Council, Barcelona Statistical Guidelines and Districts; Technical Report; Department of Statistics: Barcelona, Spain, 2012.

69 Xu, Y. Effective GPS-Based Panel Survey Sample Size for Urban Travel Behavior Studies. Ph.D. Thesis, Georgia Institute of Technology and School of Civil and Environmental Engineering, Atlanta, GA, USA, 2010.

70 Institut d'Estudis Regionals i Metropolitans de Barcelona (IERMB) and Autoritat del Transport Metropolità (ATM),Enquesta de Mobilitat Quotidiana; Technical Report; IERMB & ATM: Barcelona, Spain, 2006.

71 Cetelem PARIBAS Grupo B. N. P. The electric vehicle and the europeans. (El coche eléctrico y los europeos); Technical Report; Cetelem PARIBAS Grupo B. N. P.: Madrid, Spain, 2012.

72 Marra, F.; Yang, G.Y.; Traholt, C.; Larsen, E.; Rasmussen, C.N.; You, S. Demand profile study of battery electric vehicle under different charging options, Proceedings of the 2012 IEEE Power and Energy Society General Meeting, San Diego, CA, USA, 22–26 July 2012.

73 Valsera-Naranjo, E.; Sumper, A.; Lloret-Gallego, P.; Villafafila-Robles, R.; Sudria-Andreu, A. Electrical vehicles: State of art and issues for their connection to the network, Proceedings of the 10th International Conference on Electrical Power Quality and Utilisation, Lodz, Poland, 15–17 September 2009.

74 Load Forecasting for Low Voltage Supply; ITC-BT-10 Stanadard; Ministerio de Industria, Energía y Turismo: Madrid, Spain, 2002.

75 Red Eléctrica de España—Spanish Transmission System Operator, Available online: http://www.ree.es accessed on 7 July 2014.

76 Kostková, K.; Omelina, L.; Kyčina, P.; Jamrich, P. An introduction to load management. Electr. Power Syst. Res. 2013,95, 184–191.

77 Royal Decree (Real Decreto) 647/2011, of May 9, which regulates aggregator activities (Gestor de carga) and Time-Of-Use tariff for EV; National Law; Ministerio de Industria, Energía y Turismo: Madrid, Spain, 2011.

78 Putrus, G.A.; Suwanapingkarl, P.; Johnston, D.; Bentley, E.C.; Narayana, M. Impact of electric vehicles on power distribution networks, Proceedings of the IEEE Vehicle Power and Propulsion Conference, Dearborn, MI, USA, 7–10 September 2009.

CHAPTER 7

Control of the Electric Vehicles Power Chain with Electromagnetic Switches Reducing the Energy Consumption

Souhir Tounsi[1,2]

[1]National School of Engineers of Sfax, Sfax University, Sfax, Tunisia; [2]Laboratory of Electronics and Technologies of Information (LETI), Sfax, Tunisia.

ABSTRACT

Among the factors slowing down the production of the electric vehicles in big series, we mention the problem of weak autonomy directly bound to the weak storage capacity of the batteries. In this context, this paper describes a strategy of power chain vector control reducing the consumption and integrating a system of energy recuperation. Besides, this power chain is conceived by an analytic approach optimizing the autonomy and reducing the production cost of electric vehicle. The choice of the static converter to electromagnetic switches is a determining factor for the reliability of the global system and the reduction of the consumption. This choice poses a problem of adaptation of this low-frequency converter type to the global system that will be treated in this paper.

KEYWORDS

Electric vehicle, Motor, Converter, Control, Energy recuperation, Electromagnetic switch

1. INTRODUCTION

In look of the strong petroleum gray and the importance of the toxic gauzes clearings capacity of the thermal vehicles, the electrification of the vehicles became an actual problem. The major problems of the production of the electric vehicles in big series are the weak autonomy and the elevated cost. In this context, the whole motorconverter conceived by a systematic analytic approach, respecting the constraints of the application as the limit of the speed of the vehicle fixed to 80 km/h and interactions between the design and the control of the global system retailed in [1-3], is modelled under the Matlab/ Simulink environment, in the goal to validate this gait of design. This modelling integrates an energy recuperation system in order to reduce the consumption of the vehicles. The control strategy is vectorial imposing currents in phase with the electromotive forces, leading to an economy of the energy consumption.

2. POWER CHAIN STRUCTURE

Several configurations of power chains are presented in the literature [3]. We can mention as examples:

- The configuration with four-motor wheels to direct mechanical link or with gearing.
- The configuration to two motors in the back or in the front of vehicle to direct link or with gearing.

- The single-motor configuration with a mechanical transmission composed by a differential more gearing or without gearing. This configuration is chosen for our application since it offers the advantage of the least weak cost, since the manufacture of only one motor is less expensive than the one of several motors [4]. This configuration also permits to avoid the problem of skid, since it's impossible to control several motors to the same speed. To increase the autonomy, this structure includes a system of energy recuperation with an optimized structure.

3. TRACTION MOTOR

The motor structure is with axial flux and permanent magnet. This structure is with reduced production cost [4] (figure 1).

The d, q axis voltages are given by the following equations [5,6]:

$$V_d = R \times I_d + L_d \times \frac{\mathrm{d}I_d}{\mathrm{d}t} - \Omega \times p \times L_q \times I_q \tag{1}$$

$$V_q = R \times I_q + L_q \times \frac{\mathrm{d}I_q}{\mathrm{d}t} + \Omega \times p \times L_d \times I_d + \frac{2}{3} \times K_e \times \Omega \tag{2}$$

The electric motor constant is given by the following equation:

$$K_e = \frac{3}{2} \times N_s \times \left(\frac{D_e^2 - D_i^2}{4} \right) \times B_e \tag{3}$$

The d, q axis inductances are expressed by the following equation:

$$L_d = L_q = L - M \tag{4}$$

where L is the phase inductance [6]:

$$L = \frac{\mu_0}{2} \times \left(\frac{S_d}{e + H_a} + \frac{\left(\frac{D_e - D_i}{2} \right) \times H_d}{L_{enc}} \right) \times N_s^2 \tag{5}$$

And M is the mutual inductance [6]:

$$M = \frac{\mu_0 \times \left(\frac{D_e - D_i}{2} \right) \times \left(\frac{D_e + D_i}{2} \right) \times A_{dentm}}{2 \times (e + H_a)} N_s^2 \tag{6}$$

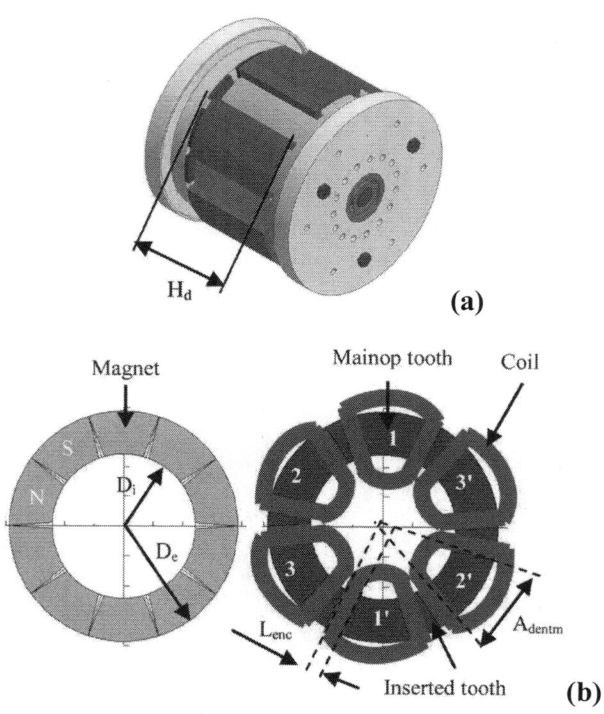

Figure 1. Motor structure. Legend: (a) 3D face of the motor; (b) 2D face of the motor.

The electromagnetic torque is expressed as follows:

$$T_{em} = K_e \times I_q + \frac{3}{2}\left(L_d - L_q\right) \times I_d \times I_q = K_e \times I_q \tag{7}$$

The electric motor model implanted under the environment of Matlab/Simulink is illustrated by the following figure 2.

4. STATIC CONVERTER WITH ELECTROMAGNETIC SWITCHES

The structure of the static converter is with electromagnetic switchs, that presents a lot of advantages in look to the structure with IGBTs [7,8]. Two structures of converter to electromagnetic switch exist to know:

- Structure with recall spring [9].
- Structure with two generating windings.

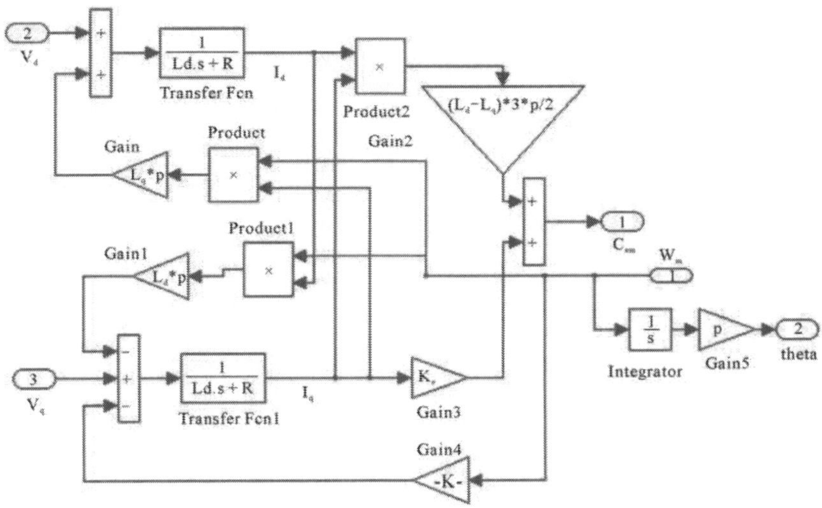

Figure 2. Block diagram of the motor implanted under the environment of Matlab/Simulink.

The structure restraint is with two generating windings, since it presents the most elevated switching frequency and the weakest maintenance cost. This structure is modular. It can include several stages in order to increase the power and the opening and closing frequency of the power contacts. A two-stage structure is illustrated by the following figure 3.

The control circuit of the converter's arm is illustrated by the following face (**Figure 4**).

The functioning sequences of the converter's arm are illustrated by the following figure 5.

The model of the static converter rests on the comparison between the references voltages come out of the currents regulators and a triangular signal with frequency f_{sw} lower then the opening and closing frequency of the electromagnetic switches (F_{ri}):

$$f_{sw} = \frac{F_{ri}}{n_{qTA}} \tag{8}$$

where n_{qTA} is the quality coefficient of phase's voltages and F_{ri} is maximal electromagnetic switches frequency:

$$F_{ri} = \frac{1}{T_r} \tag{9}$$

$$T_r = T_{off} = T_{on} = 3.74 \times 10^{-3} \tag{10}$$

The output signals of the three comparators varying to the rhythms of the reference voltages attack three hysteresis varying between U_{dc} and

$-U_{dc}$ to reproduce the pace of the food voltages delivered by the static converter.

The delay to the opening and to closing is modelled by a first order transfer function with gain equal to the unit and constant of time equal to T_r.

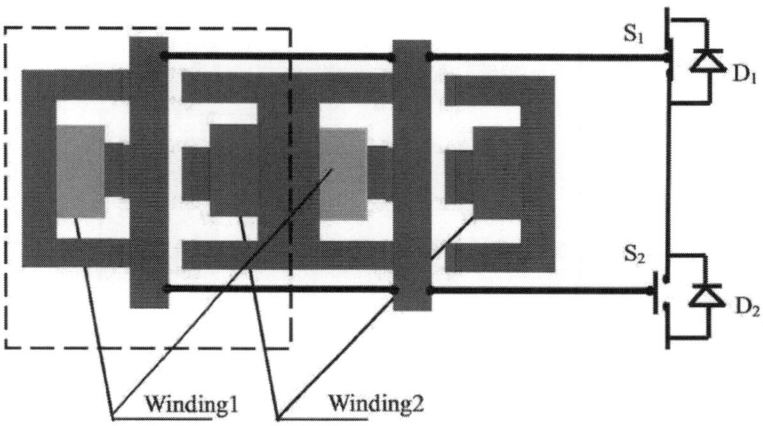

Figure 3. A two-stage structure of generating winding.

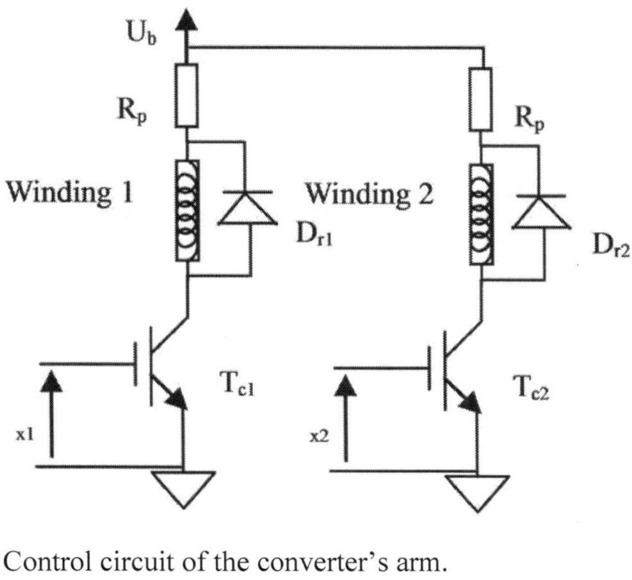

Figure 4. Control circuit of the converter's arm.

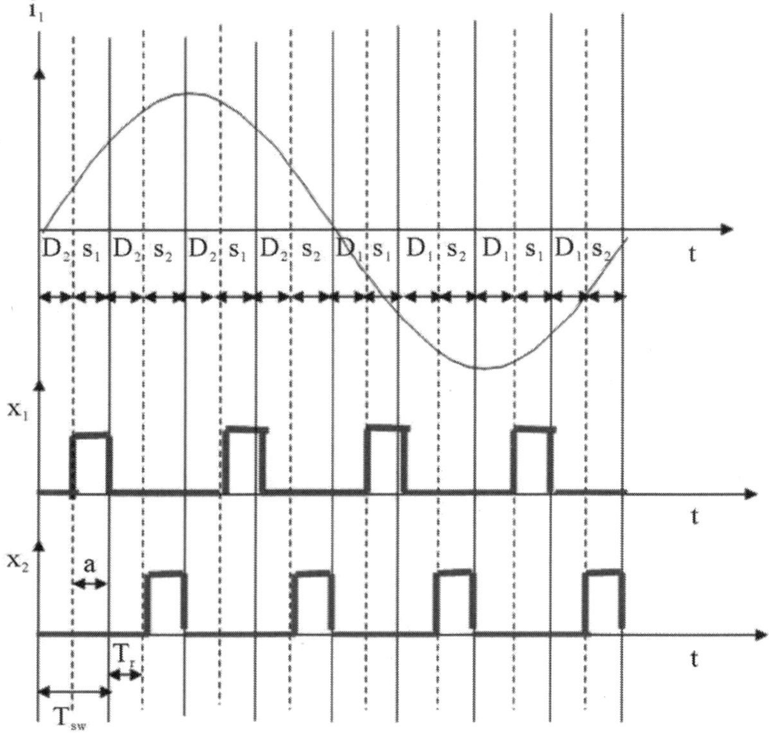

Figure 5. Functioning sequences of the converter's arm.

The converter's model with electromagnetic switches is implanted under the environment of Matlab/Simulink according to figure 6.

5. DYNAMIC EQUATION

The dynamic equation of the vehicle is deducted from the fundamental relation of the dynamics [1]:

$$\left(\frac{J \times r_d}{R_r} + M_v \times R_r\right) \times \frac{\mathrm{d}v}{\mathrm{d}t} = r_d \times C_m - \left(F_r + F_a + F_c\right) \times R_r \tag{11}$$

where:

$$F_r = f_r \times M_v \times g \tag{12}$$

$$F_a = \frac{1}{2} \times M_{va} \times C_x \times A_a \times v^2 \tag{13}$$

$$F_c = M_v \times g \times \sin(\lambda) \tag{14}$$

The insertion of a speed amplifier with gearing report r_d is in the goal to reach the maximal speed fixed to 80 km/h for our application. This report also permits to guarantee a good interpolation of the references voltages in order to have a good quality of the electromagnetic torque:

$$r_d = \frac{R_r \times F_{ri}}{n_{qTA} \times V_{max} \times p \times n_{iTR}} \tag{15}$$

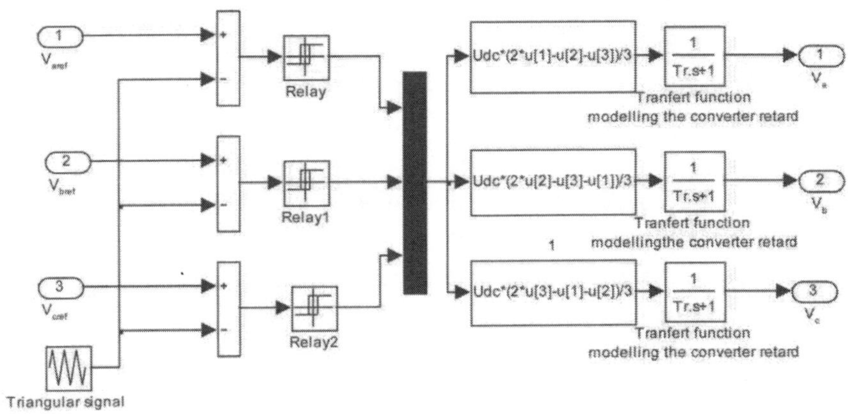

Figure 6. Model of the static converter implanted under Matlab/Simulink environment.

The dynamic equation of the vehicle is implanted under the environment of Matlab/Simulink (figure 7).

6. SPEED AND CURRENT REGULATORS

The speed regulator permits to compare the reference speed to the response speed. The output signal of the comparator attacks a proportional integral regulator (PI), serving to bring closer the response speed to the speed of reference asked by the user.

The current regulator serves to impose a current I_d equal to zero to reach the maximal value of torque (electromotive forces in phases with the currents of phase: compensation technique of the electromotive forces). The Simulink models of these two regulators are illustrated by **Figure 8**.

7. SYSTEM OF ENERGY RECUPERATION

The system of energy recuperation functions at the time of the decelerations phases corresponding to a recoverable energy. This phase of working correspond to a deceleration superior to a doorstep optimized by several simulations in order to reach the maximal recovered energy. In this phase of working, the three windings of the motor are connected to an AC-DC converter to convert the three induced electromotive forces to a continuous voltage filtered by a capacity of strong value. This voltage passes by an amplification stage (DC-DC converter with an amplification report optimized following several simulations). Finally, the amplified continuous voltage attacks the energy accumulator until the annulment of the current (The accumulator of energy is in phase of load: working in generating). When the current annuls itself, the windings of the motor connect to the converter (working in motor).

The load voltage of the energy accumulator is expressed by the following relation:

$$U_r = \frac{2}{3} \times K_e \times \Omega \times \frac{\sqrt{3}}{\pi} \times \frac{1}{1-\alpha} \tag{16}$$

To this phase of working, the torque on the tree motor T_m is expressed us follow:

$$T_m = -\frac{2}{3}\times K_e \times \Omega \times \frac{\sqrt{3}}{\pi}\times\frac{1}{1-\alpha}$$
$$\times\left(\frac{2}{3}\times K_e\times\Omega\times\frac{\sqrt{3}}{\pi}\times\frac{1}{1-\alpha}-U_{batt}\right)\times\frac{1}{R_{batt}}\times\frac{1}{\Omega} \quad (17)$$

Out this phase of working, the motor torque is expressed by the following relation:

$$T_m = K_e \times I_q \quad (18)$$

The model of the energy recuperation system implanted under the environment of Matlab/Simulink is illustrated by the following figure 9:

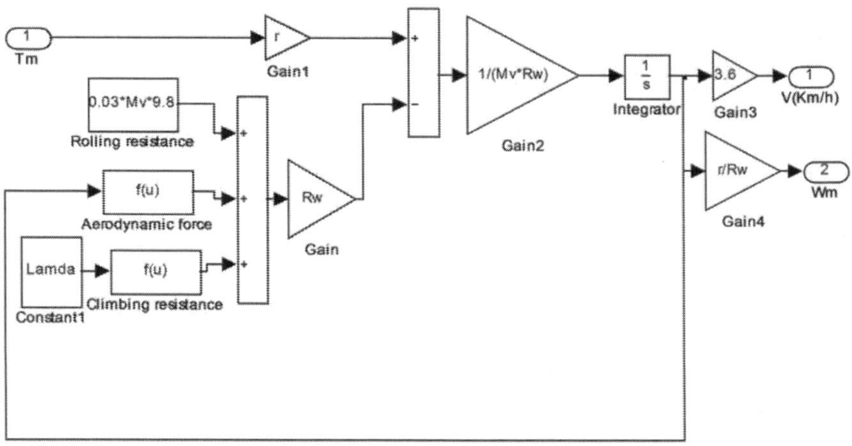

Figure 7. Dynamic equation of the vehicle implanted under the environment of Matlab/Simulink.

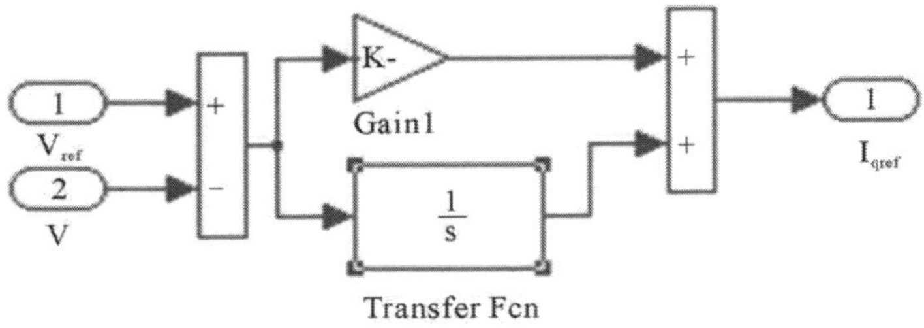

(a)

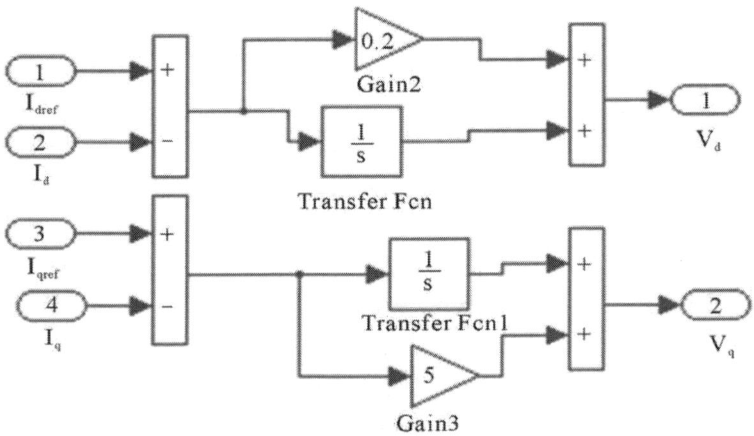

(b)

Figure 8. Simulink models of the regulators. Legend (a) Current regulator; (b) Speed regulator.

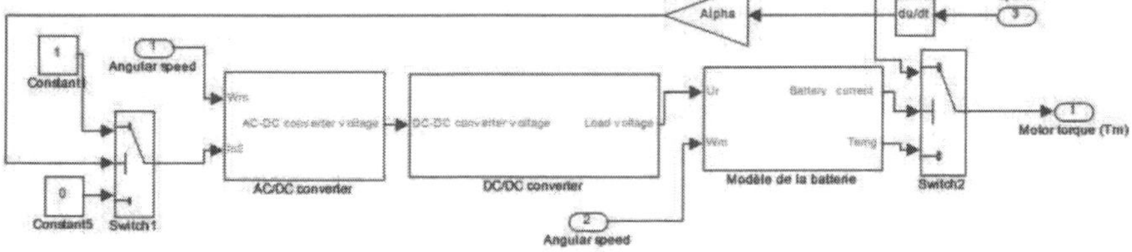

Figure 9. Model of the energy recuperation system implanted under the environment of Matlab/Simulink.

8. GLOBAL MODEL OF THE POWER CHAIN

The coupling of the different models of the electric vehicle power chain of the leads to the global model implanted under the environment of Matlab/Simulink (7.1 vesion) according to the figure 10:

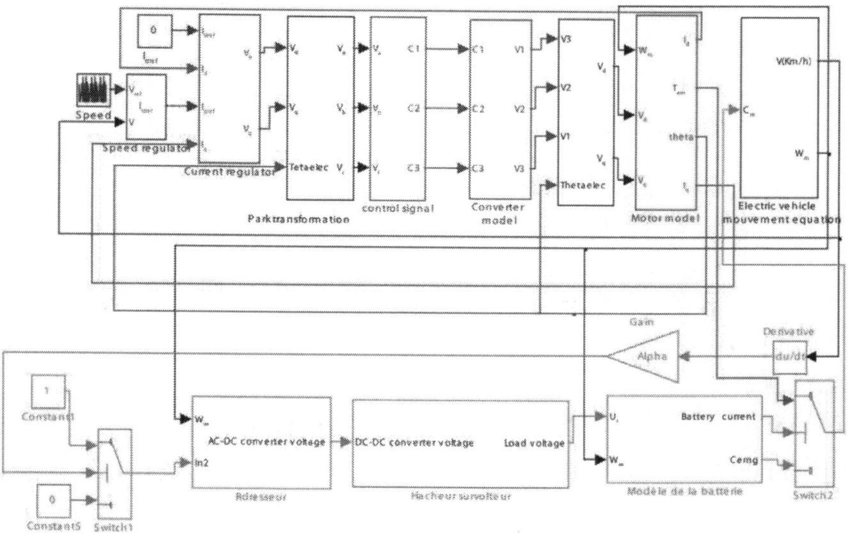

Figure 10. Block diagram of the power chain Simulink model.

The table 1 illustrates the simulation parameters of the power chain.

This model is based on the vector control technique with converter ordered by the sinus/triangle modulator. This technique, present the advantage of a reduction of the consumption on the one hand by compensation of the electromotive forces (I_d is maintained equal to zero) and by recuperation of the energy during the phases of strong decelerations on the other hand.

Table 1. Simulation parameters.

Designations	Values	Units
Electric vehicle mass (M_v)	1000	kg
Quality coefficient of phase's voltages (n_{qTA})	20	/
Switched frequency (f_{sw})	6.7	hz
Ray of the wheel (R_w)	0.26	m
Gear ratio (r_d)	1/8	/
d-q Inductances (L_d, L_q)	0.4463	mH
Phase resistance (R)	0.001	Ohm
Number of pairs poles	4	/
Back electromotive force constant (K_e)	1.125	Volt/(rad/s)
Continious voltage (U_{dc})	50	Volt

9. DESCRIPTION OF THE SIMULATION RESULTS

The speed response to an order of 80 km/h is illustrated by **Figure 11**.

This characteristic proves the importance of the mechanical time constant. This characteristic is also influenced by the parameters of the speed and current regulators, adjusted in order to slow down the response of the traction system to guarantee an acceptable use comfort (Reduction of the strong accelerations).

The evolution of the phase current according to the time for a speed order of 80 km/h is illustrated by **Figure 12**.

This characteristic shows the importance of the starting current. The current stabilizes in a weak time, what proves the reduction of the consumption at starting directly bound to the electric constant of the motor.

The pace of the electromagnetic torque for this working regime is illustrated by **Figure 13**.

This pace shows that the torque stabilizes quickly, what proves the reduction of the consumption justifying the good choice of the value of the motor electric constant. The ripple torque is as weak, what confirms the efficiency of the control technique again.

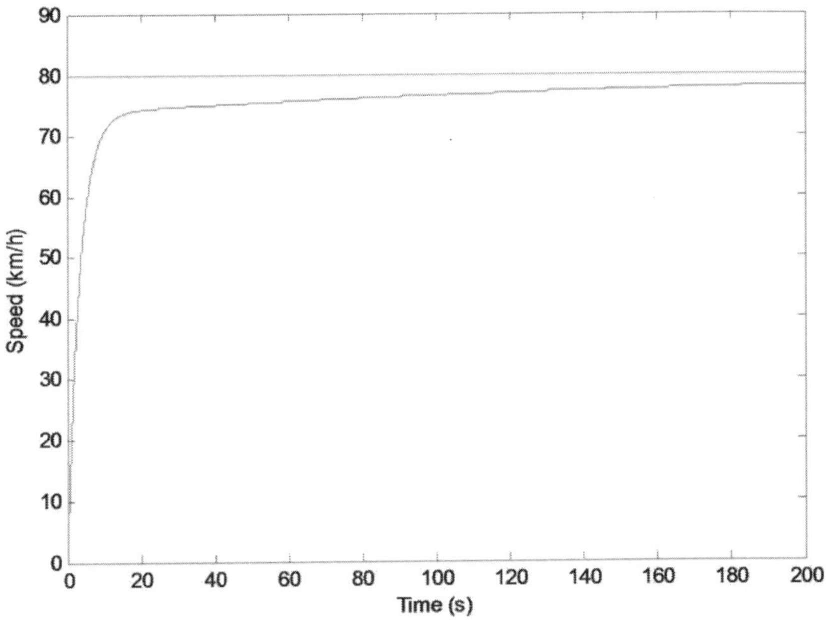

Figure 11. Speed response to an order of 80 km/h.

These paces illustrate the characteristic of low-frequency working of the power chain with electromagnetic switches. The current pace is very close to a sinusoidal shape, what shows the efficiency of the vector control technique based on the sinus-triangle pulse width modulation.

The speed response to a normalized circulation mission is illustrated by figure 15.

The speed response follows the normalized circulation mission with a good precision, what explains the efficiency of the control technique. This characteristic valid entirely the systematic design approach of the power chain retailed in [2].

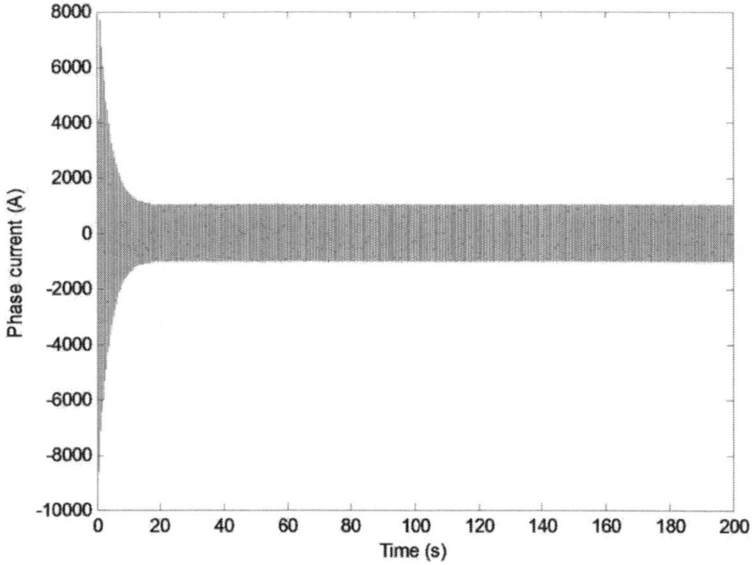

Figure 12. Evolution of the phase current according to the time for a speed order of 80 km/h.

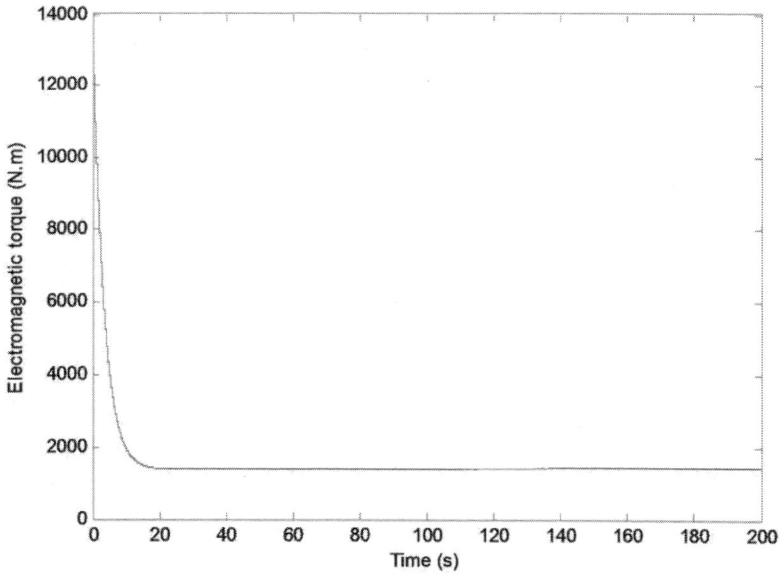

Figure 13. Pace of the electromagnetic torque for a working regime with constant speed of 80 km/h.

The pace of the current and the phase voltage are illustrated by **Figure 14**.

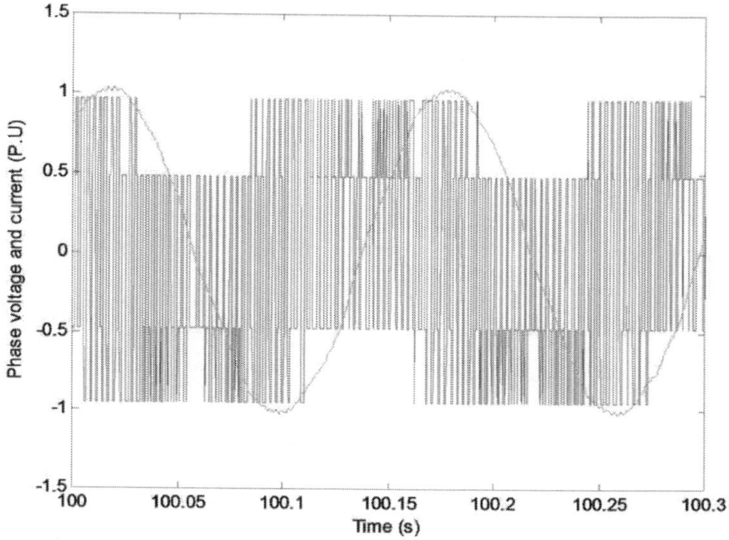

Figure 14. Pace of the current and the phase voltage for a regime of working to constant speed of 80 km/h.

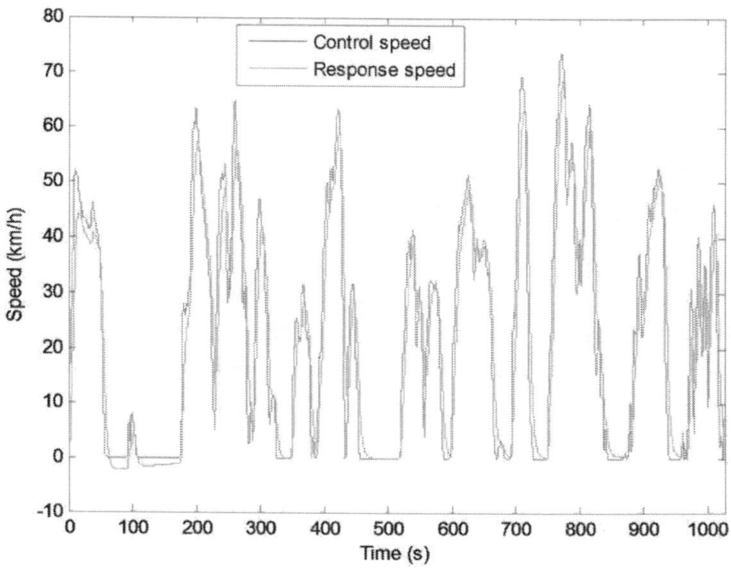

Figure 15. Response to a normalized circulation mission.

The load current of batteries (figure 16) is equal to zero during the phase of working in motor and it's different to zero during the phases of strong decelerations.

The recovered energy (figure 17) increases at the time of strong decelerations and remain constant during the phases of working in motor.

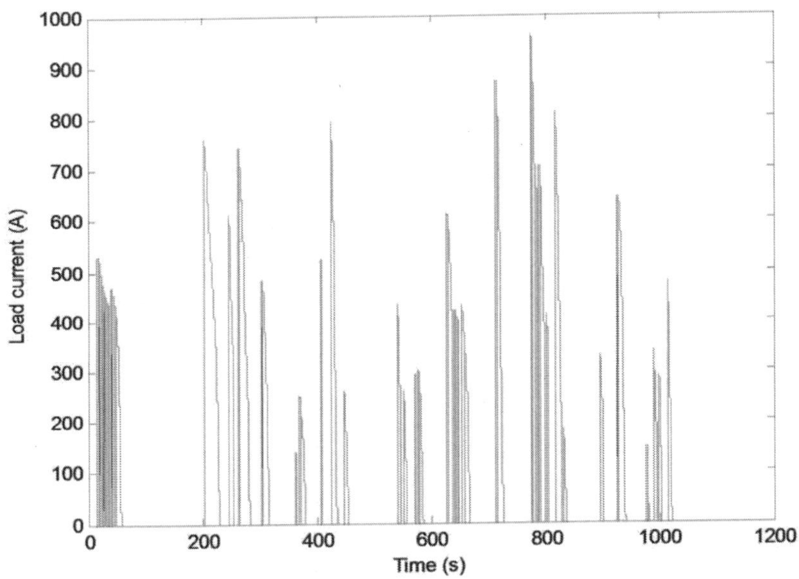

Figure 16. Battery load current.

The middle value of the recovered energy is equal to 0.0718 kwh on a length of 1027 seconds. This value is important, what shows the efficiency of the system of energy recuperation. The recovered energy improves the autonomy in relation to another system of traction without recuperation with autonomy of 120 km for a stocked energy equal to 30 kwh [10].

The torque on the motor tree (figure 18) presents negative values since at the time of working in generat ing the motor torque reverses itself.

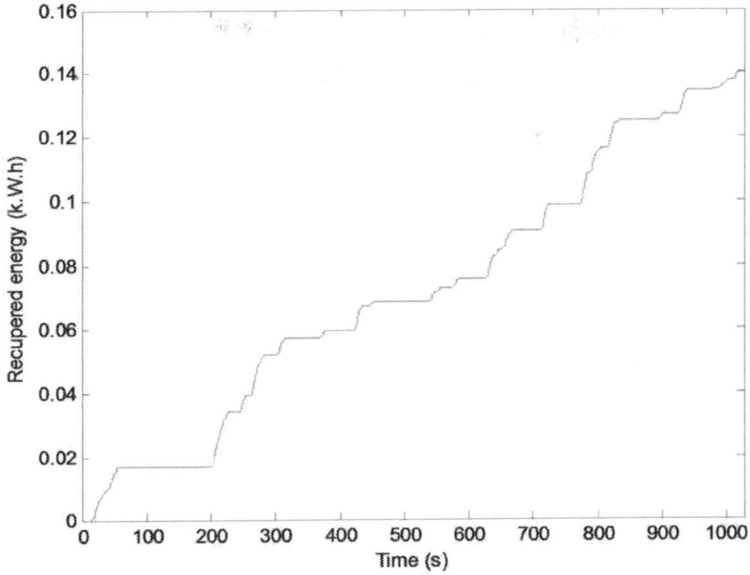

Figure 17. Recovered energy.

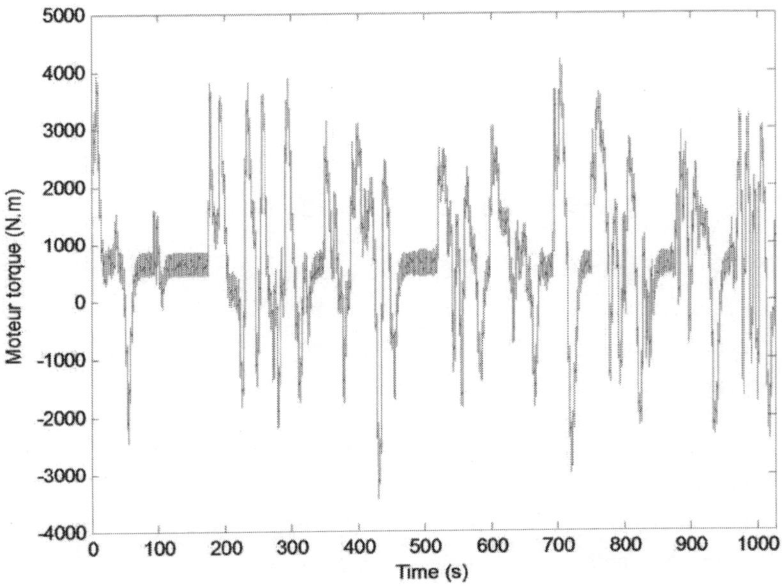

Figure 18. Torque on the motor tree.

10. CONCLUSIONS

In this paper, we present a validation methodology of the of improvement of the autonomy. Indeed, a vector control strategy integrating a system of energy recuperation optimized is developed and implanted under the environment of Matlab/Simulink. After the manufacture of the motor and the converter, it will be interesting to integrate this control law of the power chain on a DSP or FPGA to test the validity of the global design approach of this power chain reducing the consumption. Finally, the gotten results are encouraging and open the track to the industrialization of this traction system.

REFERENCES

1 S. A. Randi, S. Astier and B. Sarenti, "Full Modeling Approach of Electric Vehicle for Design Optimisation," Electric Vehicle Symposium (EVS 18), Berlin, 20-24 October 2001.

2 S. Tounsi and R. Neji, "Design of an Axial Flux Brushless DC Motor with Concentrated Winding for Electric Vehicles," Journal of Electrical Engineering, Vol. 10, 2010, pp. 134-146.

3 N. Chaker, I. B. Salah, S. Tounsi and R. Neji, "Design of Axial-Flux Motor for Traction Application," Journal of Electromagnetic Analysis and Applications, Vol. 1, No. 2, 2009, pp. 73-83.

4 R. Neji, S. Tounsi and F. Sellami, "Contribution to the Definition of a Permanent Magnet Motor with Reduced Production Cost for the Electrical Vehicle Propulsion," European Transactions on Electrical Power, Vol. 16, No. 4, 2006, pp. 437-460.

5 B. B. Salah, A. Moalla, S. Tounsi, R. Neji and F. Sellami, "Analytic Design of a Permanent Magnet Synchronous Motor Dedicated to EV Traction with a Wide Range of Speed Operation," International Review of Electrical Engineering, Vol. 3, No. 1, 2008, pp. 110-122.

6 A. Moalla, S. Tounsi and R. Neji, "Determination of Axial Flux Motor Electric Parameters by the Analytic-Finite Elements Method," Journal of Electrical Systems, Vol. 4, No. 4, 2008, pp. 398-409.

7 A. Ammous, B. Allard and H. Morel, "Transient Temperature Measurements and Modeling of IGBT's under Short Circuit," IEEE Transaction Electronic Devices, Vol. 13, No. 1, 1998, pp. 12-25.

8 M. Ayadi, A. Ammous, Y. Ounejjar and F. Sellami: "Thermal Interaction of Semiconductor Devices in Multi-Chip Modules," Systems Man and Cybernetics SMC'02, Hammamet, October 2002.

9 S. Tounsi, M. H. Kacem and R. Neji, "Design of Static Converter for Electric Traction," International Review on Modelling and Similations, Vol. 3, No. 6, 2010, pp. 1189- 1195.

10 S. Tounsi, R. Neji, N. B. Hadj and F. Sellami: "Global Optimization of Electric Vehicle Design parameters," Electric Vehicle Symposium (EVS21), Monaco, 2-6 April 2005.

CHAPTER 8

Predictive Intelligent Battery Management System to Enhance the Performance of Electric Vehicle

Mohamad Abdul-Hak[1], Nizar Al-Holou[1] and Utayba Mohammad[1]

[1] Electrical & Computer Engineering Department, University Of Detroit Mercy, Detroit, USA

1. INTRODUCTION

The Electric Vehicle (EV) is emerging as state-of-the-art technology vehicle addressing the continually pressing energy and environment concerns. The benefits of EV emerge from these vehicles' capability of sustaining their energy demands through electric grid rather than fossil fuel consumption. Well- to-Wheel studies have shown that electric drive (E-drive) offers the highest fuel efficiency and consequently the lowest emission of green house gases. Grid electricity in the United States of America has been shown to be four times cheaper than fuel given gasoline prices at $3/gallon. Consequently, it is crucial to further optimize the electric-drive mode for EV. Battery capacity should be designed to allow EV drivers reach their destination while avoiding unnecessary stops to recharge their vehicles. However, this additional battery capacity would impact the vehicle's space, weight and cost. In view of these limitations, we propose integrating EVs with the vision of Intelligent Transportation Systems (ITS). This chapter starts out by putting the design of EVs into a broader perspective by proposing a Predictive Intelligent Battery Management System (PIBMS), which will enhance the overall performance of EVs including energy consumption and emissions using the ITS infrastructure.

At the end of this chapter, the reader should have an understanding of the capabilities and limitations of the PIBMS. It lays out the design foundation for the future implementation of an interconnected EV equipped with PIBMS, which further contributes to the optimization of energy efficiency and reduced emissions.

2. EV DESIGN CHALLENGES

Recent advancement in battery and charging technologies has allowed the Electric Vehicle (EV) to be considered as the new generation of automotive transportation. However, the physical dimensions, packaging environment and charging of EV batteries continues to be the main challenge and the focus of attention in the development of EVs. Battery technology selection continues to be the primary challenge in order to achieve the proper balance in the EV design as illustrated in Figure 1 and described below.

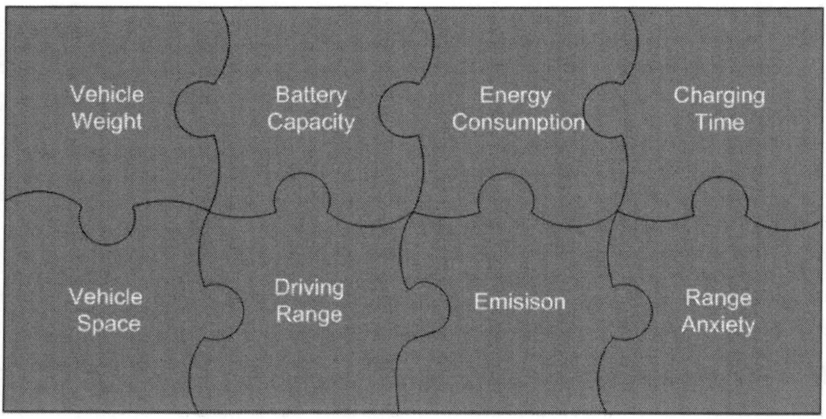

Figure 1. EV Design Parameters

- Battery Capacity: EV battery capacity is predetermined by the battery design and cell chemistry. Lithium polymer batteries are the target implementation for EV due mainly to their high power-to-weight ratio.

- Vehicle Weight: EVs weight increases proportionally to battery capacity increase.

- Vehicle Space: Vehicle operators favor personal use of vehicle space. EV requires more packaging space to house the battery in a safe environment. Generally, the battery is packaged in the center of the vehicle where vehicle operators conventionally utilize this space.

- Driving Range: EV can only run for 100-200 miles before recharging compared to gasoline vehicle, which can drive more than 300 miles before refueling.

- Charging time: EV has no internal source for recharging the battery. EV charging time ranges between 3-8 hours compared to 2-4 minutes of refueling for gasoline vehicle.

- Range Anxiety: EV operators are usually concerned with their vehicles' limited driving range, inadequate charging infrastructure, and long charging time.

- Energy Consumption: EV propulsion systems offer around 85% efficiency compared to about 25 % efficiency for Internal Combustion Engines (ICE).

- Emission: EV emits no pollutants; however, power plant generating the EV electricity may emit them.

While battery manufacturers are still pursuing further improvement in energy capacity, the navigation technology and rapid advances in wireless communication technology can be used to achieve the vehicle performance balance described as "Target" and presented in Table 1.

Table 1 clearly shows the limitations for utilizing battery capacity as the only design variable for achieving a balanced EV design that is acceptable for EV operators. To realize the success of EVs, achieving the "Target" design option shall be exerted. This call for considering two crucial aspects in addition to battery technology: Traffic information and wireless communication.

The need to identify traffic conditions and the ability to transfer these conditions constitutes the success of optimizing energy consumption and emission reduction in EVs.

Realizing emission reduction in EVs is a crucial step toward emission free vehicles. However, it requires a better understanding of emission free EVs and the ensuing energy sources. These topics are addressed in the following section: EV Emission.

Table 1. EV System Design Option Evaluation

	Design Options		
Parameter	A	B	"Target"
Battery Capacity	↗	↙	↗
Vehicle Weight	↗	↙	↙
Vehicle Space	↙	↗	↗
Driving Range	↗	↙	↗
Charging Time	↗	↙	↙
Range Anxiety	↙	↗	↙
Energy Consumption	↗	↙	↙
Emission	↗	↙	↙

→ Positive Impact
→ Negative Impact

2. EV EMISSION

2.1. Electrical Energy Source

An accurate assessment of EV emission requires the inclusion of the electrical energy source associated emission with the generation and transmission. Electrical energy is generated from two main sources as illustrated in Figure 2:

- Non-renewable source: Coal, natural gas, nuclear, petroleum
- Renewable source: wind, solar and geothermal

Non-renewable energy produces elevated Greenhouse Gas (GHG) emissions. Coal is leading all other energy sources in terms of GHG emissions.

Renewable energy investment has to some extent been very limited due to the associated high development cost. However government subsidiaries continue to make the renewable energy investment more affordable.

The claim of EV proponents is that this type of vehicle is a Zero Emission Vehicle (ZEV). This claim depends on many factors, but a key factor shall be highlighted. The EV operating energy emission is a function of the energy source. The upstream Greenhouse gases (GHG) emission is based on power plant types and efficiency. The claim of EV technology proponents that this type of propulsion technology will offer a potential to reduce a long-term GHG emission can be verified with the implementation of the Well-To-Wheel (WTW) emission model to analyze the GHG emission of the electrical energy source.

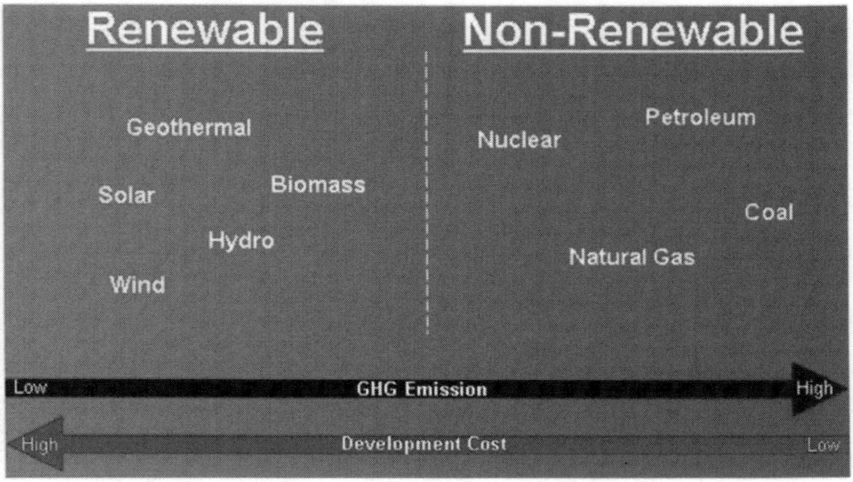

Figure 2. Electrical Energy Source

2.2. Ev Configuration

The EV is mainly a conventional vehicle with the following main differences illustrated in Figure 3and listed below:

1. High voltage electric battery rather than a fuel tank to store and supply the required operational energy
2. Electric motor rather than an internal combustion engine to propel the vehicle
3. Gear box rather than a transmission to couple the power from the electric motor to the drive shaft
4. On Board or Off Board Charger to allow for recharging of the high voltage electric battery
5. Direct current / Alternating current (DC/AC) inverter to convert the DC high voltage battery into AC to drive the E-motor
6. DC/DC converter to convert the DC high voltage battery into DC low voltage battery (Conventionally identified as 12 Volt battery)

2.3. Ev Efficiency

The EV overall efficiency can be classified in three main categories. The following section describes the categories and their respective components:

2.3.1. Charging Efficiency

Automotive charging standards are currently being developed worldwide to allow for DC (Direct Current) charging. In contrast AC/DC (Alternating Current / Direct Current) charging standards have already been established and are currently being implemented in a number of alternative vehicle technology production models such as the Chevy Volt, EV SMART, Mitsubishi MIEV, Nissan Leaf and Tesla. DC charging enables the vehicle's high voltage DC battery to be directly charged from the charge station bypassing the vehicles' on board charger thus further improving charging efficiency and time. DC charging is the target implementation for public charging enabling fast charge. Due to the associated high cost of DC charging infrastructure, AC/DC charging will be the alternative and only solution for residential charging.

The EV charging efficiency is the ratio of energy transferred to the high voltage battery to the energy consumed from the AC source. Charging efficiency is highly dependent on charging power and operating temperature. Figure 4 depicts a typical EV charging efficiency operated at room temperature and utilizing an AC/DC onboard charger with a maximum output power of 3500 W.

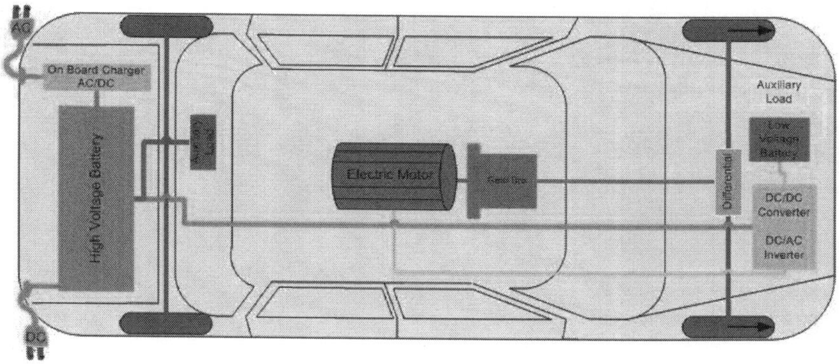

Figure 3. Electric Vehicle Model

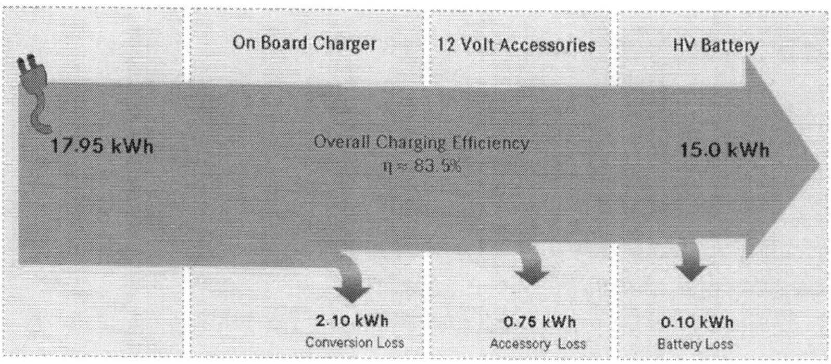

Figure 4. EV Charging Energy Flow and Efficiency Diagram

2.3.2. Operational Efficiency

Generally the efficiency of the EV Electrical Motor (EM) is exceptionally high ~ 85 % compared with an Internal Combustion Engine (ICE) ~ 25 %. Power losses in an EV are negligible, in this section we will focus on power losses from key components occurring in an electrical propulsion system during driving mode due to power conversion, operation and propulsion. As illustrated in Figure 5approximately 81.3 % of the energy stored in the HV battery is utilized to propel the EV. Combining the EV overall charging efficiency with the EV overall operational efficiency, the EV efficiency becomes ~ 67.9 % around four times more efficient than an ICE propelled vehicle with an overall efficiency of ~ 14 %.

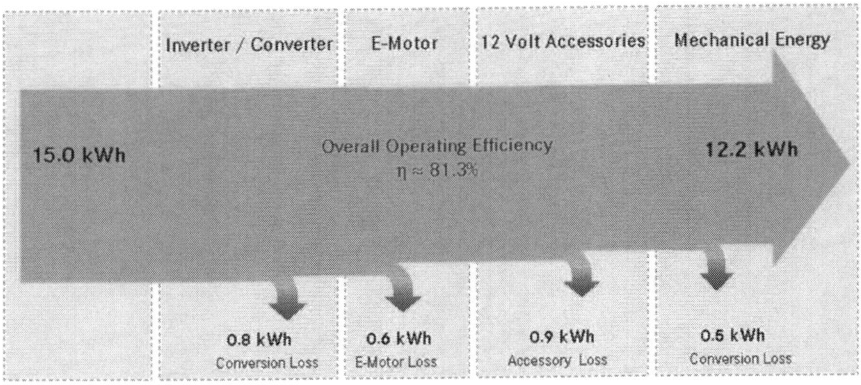

Figure 5. EV Operating Energy Flow and Efficiency Diagram

2.3.3. Power Source Generation And Transmission Efficiency

For a full representation of energy and emission calculation, it is important to consider the efficiency involved in energy recovery, processing and transportation. Complete vehicle energy-cycle analysis tools, commonly known as Well-To-Wheel (WTW) analysis tools are needed to provide an accurate assessment of EV overall efficiency and emission.

The U.S Environmental Protection Agency's (EPA) offers an emission database "National Emissions Inventory" (NEI); the database includes annual emissions associated with electric energy generation.

To fully evaluate emission impact of EV, a Well-to-Wheel emission model shall be considered such as the Greenhouse Gases, Regulated Emissions, and Energy Use in Transportation (GREET). GREET was developed by the U.S. Department of Energy (U.S. DOE) to allow researchers to evaluate emissions from a full fuel cycle for EVs and other various propulsion technologies as depicted in Figure 6.

3. PIBMS ARCHITECTURE

Current EV architecture incorporates a Battery Management System (BMS) which is a vehicle integrated electrical module responsible for monitoring State Of Charge (SOC) and maintaining a suitable state of health (SOH) of the EV high voltage battery through controlled charging and discharging processes of the battery cells.

Figure 6. WTW Emission Modeling

Adding a predictive and an intelligent component to the BMS design can make the architecture of the EV more energy and emission efficient, as it would facilitate acquiring traffic condition; offer a dynamic response to all future stochastic traffic flow situations through travel route and drive profile advisement. The integration of the predictive and intelligent components with the BMS led to the concept of the Predictive Intelligent Battery Management System (PIBMS).

With the rapid advances in wireless communication, global positioning system and the introduction of smartphones, the world is transitioning form being an online connected world to become a mobile connected world. The PIBMS concept will be based on further developing and integrating the existing technologies of the Global Positioning System (GPS), the wireless communication technology "Dedicated Short Range communication" (DSRC), and the advanced computing mobile phones identified as smartphone. The PIBMS receives traffic information from traffic light controllers and roadside units, location data from GPS and charge point data in the vicinity as depicted in Figure 7.

3.1. Intelligent Transportation System (Its)

The technological progress in wireless communication, Global Positioning System (GPS) and vehicle electronics is enabling the introduction of advanced technologies into the transportation system commonly referred to as the Intelligent Transportation system (ITS).

3.1.1. Dedicated Short Range Communication (Dsrc)

Dedicated Short Range Communication (DSRC) defined by the framework of the international standards organization ASTM and standardized by the IEEE 802.11, IEEE P1609.x and SAE J2735 standards, is a two-way short- to medium range (~1000 meters) wireless communication designed for automotive application and currently being systematically deployed throughout the U.S transportation system across the nation and sponsored by the U.S Department of Transportation Research and Innovative Technology Administration (RITA). DSRC enables the attainment of the following vehicle safety critical components for vehicle communication:

- Fast Network Acquisition: To allow immediate establishment of communication between vehicles and road side units
- Low Latency: To allow least execution time
- High Reliability: To allow high level of user reliability

DSRC offers the base and single wireless communication technology for future vehicular safety communications; furthermore DSRC is gaining momentum among researchers for future vehicular applications focused on energy optimization and emission reductions.

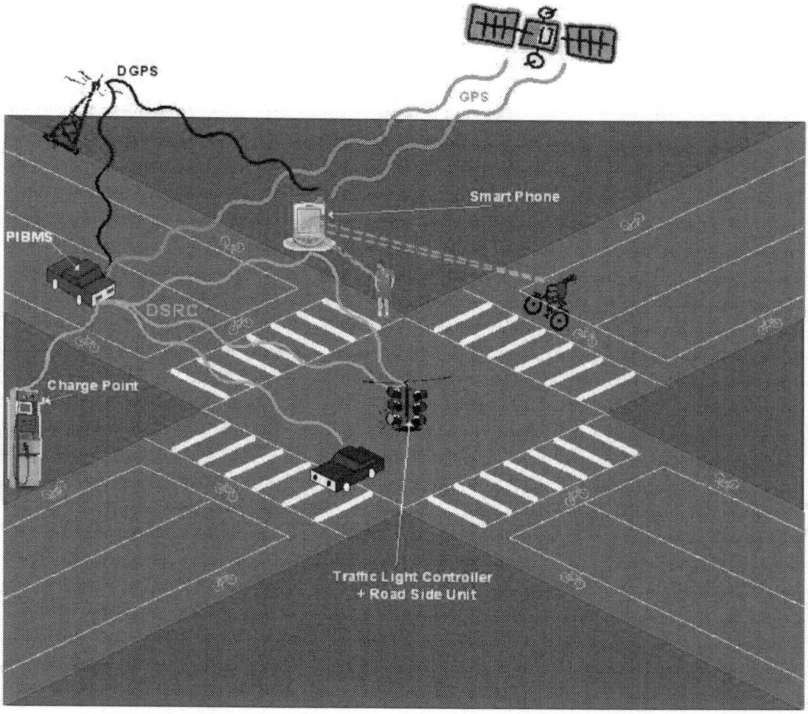

Figure 7. PIBMS Communication Block Diagram

3.1.2. Positioning System

Real time vehicle position is required with a very high level of accuracy; this would allow the system to optimize the output with both high confidence and reliability.

Several satellite receivers' manufacturers offer systems with an extremely superior accuracy such as the Topcon GR-3 receiver. The Topcon GR-3 is compatible with the US satellite system GPS, the Russian satellite system GLONASS and the European satellite system

GALILEO. This receiver system claims a static accuracy of 3mm, a Real-Time Kinetic (RTK) accuracy of 10 mm.

It is important to note than in cases where satellite navigation coverage is not available due to for example driving through tunnels, Differential GPS data shall be used which would offer in this case a slightly reduced accuracy.

3.2. Pibms Design

The PIBMS distinctive features are the predictive and the intelligent:

1. The predictive feature of the PIBMS is viable through the capability of the EV to integrate the capability of vehicular wireless communication technology (DSRC) to communicate traffic, charging infrastructure and vehicle data.

2. The intelligent feature is offered through the application of a 2-scale dynamic programming optimization approach onboard the PIBMS operated EV. The proposed PIBMS architecture consists of six modules illustrated in Figure 8 and described below:

3. Traffic Data Extractor (TDE): To extract the future traffic data from the ITS network including traffic flow, intersection light status. This data is consequently used to determine if alternative routes should be considered.

4. Vehicle Operation Mode (VOM): To provide the vehicle's current operation modes including vehicle speed, gear and SOC.

5. Trip Model Identifier (TMI): To learn the route road condition including slope and distance, this is accomplished through the use of GPS data.

6. Trip Model Deflector (TMD): To re-route trip as found necessary following the processing of future traffic data.

7. State Of Charge Optimizer (SOCO): To optimize battery energy the intelligent algorithm found in this module is expected to have two key criteria: agile and dynamic.

8. Driver Feedback Control (DFC): To provide the driver feedback relative to drive style including speed, acceleration, and deceleration.

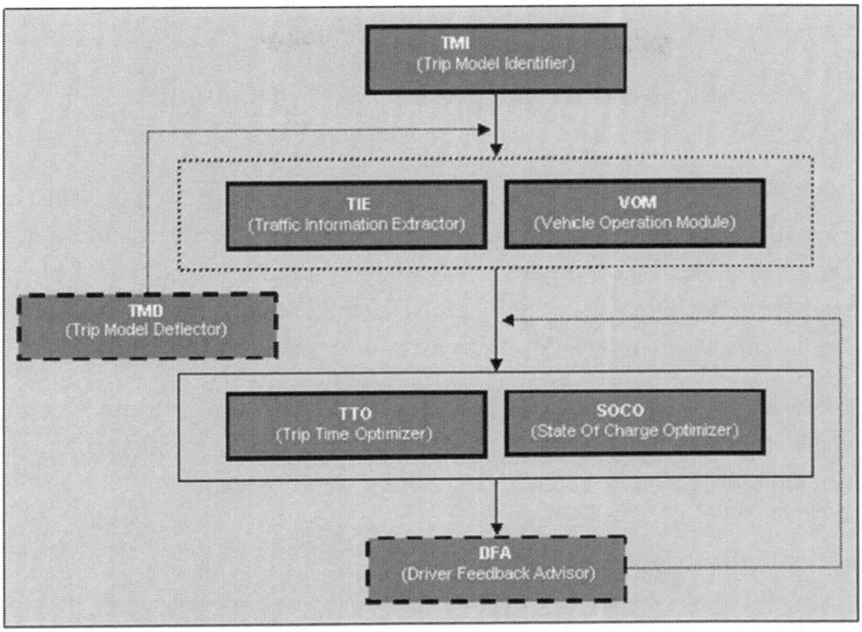

Figure 8. PIBMS Architecture

3.2.1. Design Realization

The system design is to offer the user the ability to follow advised upon route model, trip model and electrical accessory model to optimize the following queries:

1. Energy consumption
2. Emission
3. Travel Time

The system design consists of three phases:

- Phase I uses initial vehicle and route parameters at origin to advise the driver for an initial drive and route profiles based on the historical traffic data.

- Phase II uses the driver input to arrange the cost function to be optimized based on driver selective or combined selection of optimization criteria: energy consumption, vehicle emission and travel time.
- Phase III uses energy, emission and travel time cost function to plan optimized route, drive and accessories profiles.

The PIBMS in combination with the Intelligent Transportation Systems (ITS) data offers the capability of operating the EV amid additional energy optimization and emission reduction. Figure 9illustrates a block diagram of the Predictive Intelligent Battery Management Sub-Systems and the interfaces to vehicle sub-systems and road infrastructure.

The historical traffic data is utilized for initialization of the system, where vehicle has not yet established a real-time connection with the infrastructure and other vehicle's communication system.

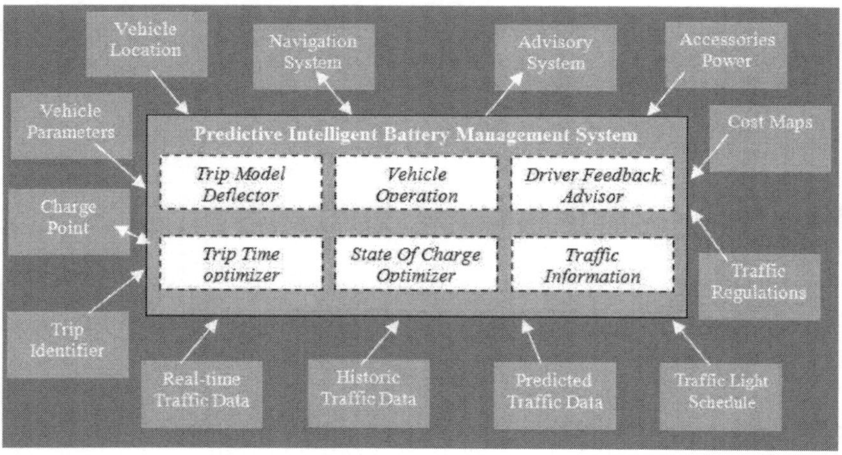

Figure 9. PIBMS Interface Block Diagram

Referring now to Figure 10, the PIBMS is first initialized with an array of input data including: vehicle position, destination, and the maximum allowed arrival time as set by the driver. The PIBMS in coordination with the navigation system will generate an energy and emission efficient route and drive profile based on historical traffic data. The route will be established in multiple segments. The PIBMS will determine the number of route segments and assign a maximum constraint in terms of consumed time, energy, and emission to travel those N segments. The PIBMS selects the Kth sequence associated with the expected segment to be traveled next. Through current and predicted traffic data, the PIBMS again in coordination with the navigation system will iteratively seek, compute and compare energy, emission and travel time of the selected segment. If the segment meets or is inferior to original estimated time, emission and energy consumption target costs, driver is advised to maintain driving through the formerly selected segment with its associated driving profile. Lest the originally selected segment does not meet the energy, emission and time constraints, an alternative segment is inquired and projected by the navigation system as an advisory path and drive profile. The navigation system will continuously seek alternative segments; if none are obtained the PIBMS will select the segment with the least calculated cost function. The driver is then advised to follow the recommended segment and driving profile to reduce energy consumption and emissions. In all cases, when driving and route profiles are established, they will be communicated to the Road Side Unit (RSU) to be then utilized as predictive traffic data for other PIBMS equipped vehicles seeking their respective optimized drive and route profiles, in addition to pre-scheduling arrival time to charge point as necessary. It is very critical for EV operators not to be abandoned mid-route where vehicle have no other self propelling means, an alternative route offering re-charge station is recommended to the driver. Provided that final segment has been computed and selected, thus indicating that driver is to commence driving the last segment of the route thus arriving to destination.

3.2.1.1. Phase I: Initialization

At initialization the system calculates a drive and route profiles based on historical traffic data provided by the roadside unit server. The historical traffic data includes current traffic flow conditions for the route profile from origin to destination in addition to charge point locations and energy source emission.

3.2.1.2. Phase II: Optimization

In this phase the route is divided into segments, for each possible path in the segment the travel time, energy consumption and emission are calculated in advance.

Cost Function

The parameters to be selectively or in combination optimized are travel time T(i) energy δ(i), and emission E(i), thus the cost function to be minimized can be stated as follow:

$$C = \sum_{i=0}^{j-1} \left[\alpha . t(i) + \beta . \delta(i) + \omega . E(i) \right] = \sum_{i=0}^{j-1} TC\left[x(i), u(i) \right] \tag{1}$$

where j is the period of the route, α, β and ω are the respective weights for travel time, energy consumption and energy source emission. The PIBMS is to allow the driver to select at initialization the priority of travel time, energy consumption and energy source emission through associating the weights α ; β ; ω, TC is total cost of route as well as energy, emission and time, x (i) is the dynamic state vector of the vehicle such as vehicle speed, energy capacity, motor torque, vehicle route and u (i) is the control vector of the vehicle such as the recommended vehicle speed, the recommenced acceleration, the recommended deceleration, the recommended route and the recommended charge point. The optimization problem becomes the search for the control vector u (i).

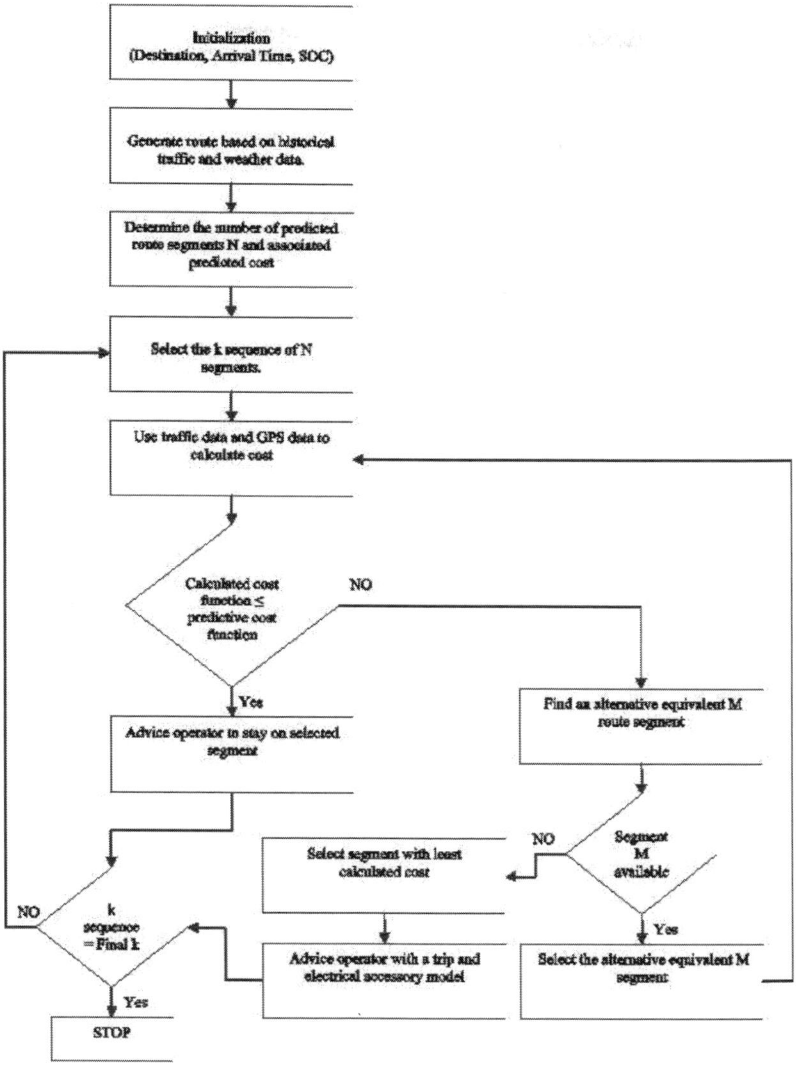

Figure 10. PIBMS Operational Flow chart Diagram

The cost function will be subject to the targeted EV model thus resulting in the following constraints:

$$
\begin{aligned}
&\delta_{min} \leq \delta\ (i) \geq \delta_{max} \\
&E_{min} \leq \delta\ (i) \geq E_{max} \\
&T_{min} \leq T\ (i) \geq T_{max} \\
&V_v(i) = V_{v\text{-}req} \\
&\tau_{em}(i) = \tau_{em\text{-}req}(i) \\
&V_v(i) \leq V_{v\text{-}max} \\
&V_{minlimit} \leq V(i) \geq V_{maxlimit} \\
&Acc_{v\text{-}min} \leq Acc_v(i) \geq Acc_{v\text{-}max} \\
&\tau_{em}(i) \leq \tau_{em\text{-}max}(i)
\end{aligned}
\tag{2}
$$

where V_v is vehicle velocity, $V_{v\text{-}req}$ is requested vehicle velocity, $V_{v\text{-}max}$ is vehicle maximum velocity, $V_{minlimit}$ is minimum speed limit, $V_{maxlimit}$ is maximum speed limit, $Acc_{v\text{-}min}$ is vehicle minimum acceleration, $Acc_{v\text{-}max}$ is vehicle maximum acceleration, τ_{em} is electric motor torque and $\tau_{em\text{-}req}$ is electric motor torque requested, $\tau_{em\text{-}max}$ is electric motor torque maximum.

In this phase a modified 2-scale predictive dynamic programming approach is implemented for obtaining the optimal drive and route solutions. Route is divided into segments, for each possible path in the segment the energy consumption and emission are calculated. The 2-scale predictive dynamic programming approach is implemented based on Bellman's Principle of Optimality; the optimal solution is obtained by identifying the initial and terminal conditions of the state. The optimized solution is then for segment k ($0 < k < i-1$) realized by minimizing the cost function defined as:

$$C_i^*[x(i)] = \min_{u(i)} \left\{ TC[x(i), u(i)] + \left[C_{x+1}^*[x(i+1)] \right] \right\} \tag{3}$$

The algorithm shows how to plan optimized drive and route profile based on energy, emission and travel time. The PIBMS will generate advisory audible/visual recommendations to the driver, including optimized routes, optimized drive profiles, and optimized electrical load profiles. The optimization algorithms will take into consideration improving traffic safety, congestion, energy consumption, emission and travel time.

The current and predicted traffic conditions are maintained by the RSU. In order to reduce bandwidth utilization for communication between RSU and PIBMS, repeating traffic data will be communicated to PIBMS only in case of traffic condition change thus eliminating message overhead.

4. SIMULATION ENVIRONMENT

To evaluate the performance of the PIBMS a simulation tool integrating traffic, vehicle and network models shall be employed. The simulation platform PIBMST (Predictive Intelligent Battery Management Simulation Tool) consists of three main building blocks described below and illustrated in Figure 11 The traffic, vehicle and network models are to allow for bidirectional coupling among each other, thus resulting in enhanced simulation accuracy however at the cost of increased computation time.

4.1. Traffic Model

A recent study has proven that the traffic mobility model can be represented by trace-based models based on generated vehicular traces. The traces are real world based on mapping the positions of the vehicles. The traffic models are in general classified in two categories:

- Macroscopic: Mathematical model to simulate major traffic characteristics such as average speed, density, and flow.
- Microscopic: In contrast to the macroscopic mathematical model, the microscopic model is to simulate single vehicle traffic characteristics.

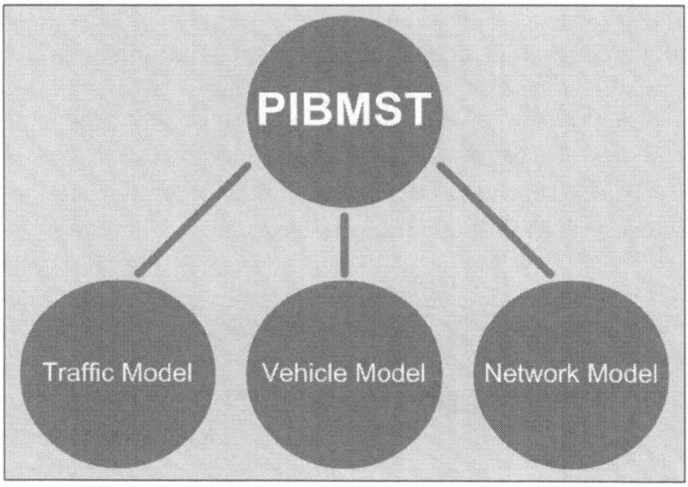

Figure 11. PIBMST Platform

Traffic flow is a dynamic problem, influenced by several variables thus for any beneficial performance of the trip model the microscopic traffic model is implemented in PIBMST.

4.2. Vehicle Model

The EV model represents a series of mathematical equations representing the characteristics of the identified EV components in Figure 3 and the forces applied to the vehicle as depicted in Figure 12

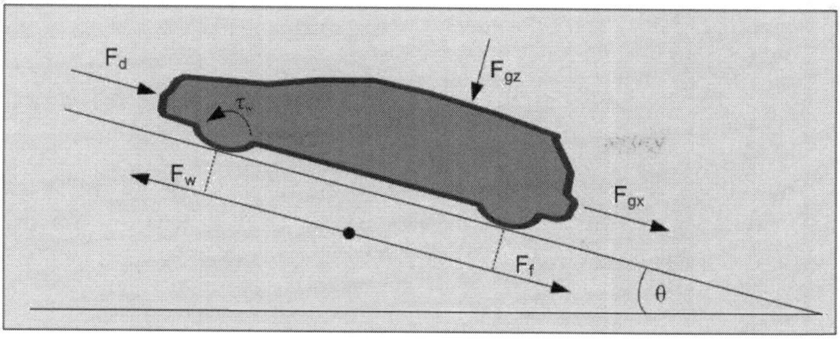

Figure 12. Applied Forces to the EV

The earth's gravitational force imposes a force F_g on the vehicle. F_g is derived from Newton's second law where a body of mass m endure an acceleration a resulting in an applied net force F.

$$F = mg \tag{4}$$

The gravitational normal force applied to the vehicle shall take into consideration the slope angle θ, when vehicle is moving uphill or downhill.

$$F_{gz} = mgCos\theta \tag{5}$$

$$F_{gx} = mgSin\theta \tag{6}$$

In order to move the vehicle a wheel force F_w is applied on the wheel. F_w is the resulting force from the generated torque in the electric motor applied to the vehicle's wheels through a gear box with a fixed differential ratio. F_w is then represented as the ratio of the torque applied to the wheel τ_w to the wheel radius, r_w.

$$F_w = \frac{\tau_w}{r_w} \quad (7)$$

When vehicle is moving the aerodynamic drag force F_d is created. F_d depends on the air density ρ, the vehicle frontal area A_v, the drag coefficient C_d, and the vehicle velocity V_v.

$$F_d = \frac{1}{2} C_d \rho V_v^2 A_v \quad (8)$$

The contact surfaces between the vehicle's wheels and the road results into a friction force F_f. The product of the friction coefficient μ_f and the vehicle's gravitational force F_g results in the corresponding frictional force F_f.

$$F_f = \mu_f F_{gz} \quad (9)$$

The total force acting on the vehicle F_t is the sum of all applied forces on the vehicle in the driving direction.

$$F_t = F_w - F_{gx} - F_d - F_f \quad (10)$$

The acceleration of the vehicle is determined by the torque applied to the wheels. The wheel torque $\tau_w$is the product of the E-motor torque τ_{em} the gear box ratio G_b

$$\tau_w = \tau_{em} G_b \quad (11)$$

The acceleration of the vehicle a_v through the application of Newton's second law is the ratio of the total force acting on the vehicle to the mass of the vehicle

$$F_t = \frac{F_w - F_{gx} - F_d - F_f}{m}$$
$$= \frac{\tau_w}{r_w m} - gSin\theta - \frac{1}{2m} C_d \rho V_v^2 A_v - \mu_f g Cos\phi \quad (12)$$
$$= \frac{\tau_{em} G_b G_i}{r_w m} - gSin\theta - \frac{1}{2m} C_d \rho V_v^2 A_v - \mu_f g Cos\phi$$

The angular velocity of the E-motor ω_{em} is the angular velocity in rotation per minute (RPM) multiplied by the E-motor turnover rate 2 π and divided by 60 (to transform RPM into revolution per second)

$$\omega_w = \frac{2\prod \omega_{em}}{60Gb} \quad (13)$$

The vehicle speed is the product of the wheel radius and the angular velocity of the wheel. rotation per minute (RPM) multiplied by the E-motor turnover rate 2 π and divided by 60

$$Vv = r_w \omega_w = \frac{r_w 2\prod \omega_{em}}{60Gb} \quad (14)$$

4.2.1. Emission Model

The emission model considers the emission associated with the generation and transportation of electricity in addition to the operation of the EV as illustrated in Figure 13.

The EV emission model is to be based on governmental accredited agencies such as the U.S. Environmental Protection Agency's (EPA's) electric power plant emission database. The EV emission is the product of consumed electrical energy in Kilo Watt hour (KWh) and the associated emission of the electrical energy source and transmission to the EV in grams (g) per KWh according to EPA. The results are presented in g/KWh of VOC, CO, CO_2, NO_X, PM_{10} and SO_X.

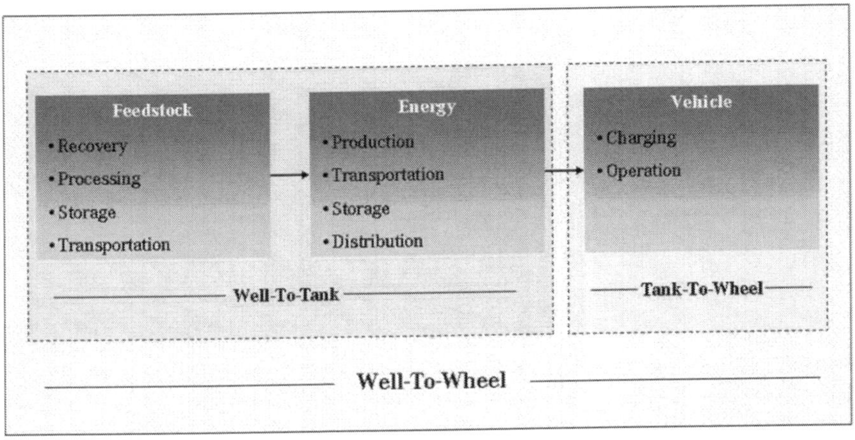

Figure 13. Well-To-Wheel Emission Analysis Model

4.3. Network Model

The roadway includes dynamic nodes such as vehicles, cyclists and pedestrians, and the static nodes such as Road Side Unit, Traffic Light Controller and Charge Point. The simulation of the nodes will require the implementation of a Vehicular ad-hoc network (VANET) capable of simulating the behaviour of the DSRC network. The network data model simulation is a discrete event simulator, implementing the protocol stack Wireless Access in Vehicular Environments (WAVE)/ Dedicated Short Range Communication (DSRC) as illustrated in Figure 14.

5. CONCLUSION

Due to the single propulsion system design in the EV, the latter offers the consumers a greater reliability, simplicity of maintenance and vehicle cost compared to Plug-In Hybrid Electric Vehicle (PHEV). Further more compared with the Fuel Cell Vehicle (FCV) the EV is more advantageous relative to vehicle cost, recharging infrastructure and safety.

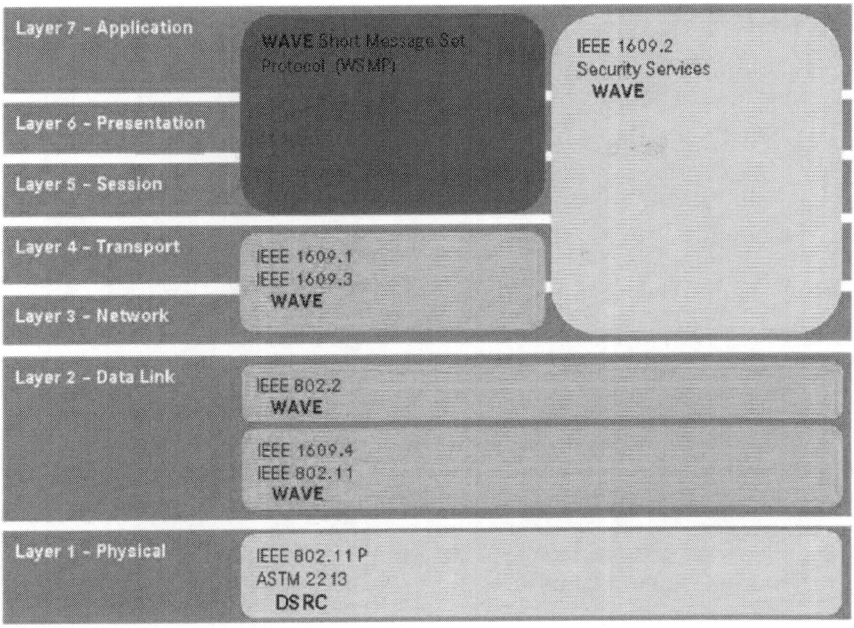

Figure 14. Protocol Stack

The automotive industry is being reshaped with the development of the EV. The new generation of automobiles are demanded to meet the market's conventional demands from vehicle space, driving range and convenience; furthermore new requirements have been shaped by the market to include energy consumption and environmental impact.

The EV will lead the way among the alternative vehicle technologies to target energy consumption and emission reduction.

This chapter offered the conceptual framework for the PIBMS application using DSRC and GPS technologies to offer EV operators an enhanced energy efficiency and decreased emission. Furthermore the proposed framework is designed to target near future implementation for relatively negligible cost using existing equipments and technologies.

REFERENCES

1. Martin Eberhard and Marc Tarpenning,2006The 21st Century Electric Car," Tesla Motors Inc, July 19, 2006

2. J. J. Romm, A. A. Frank, 2006Hybrid Vehicles Gain Traction,Scientific American294n4, April 2006,763770

3. M. Abdul-Hak, N. Al-Holou, 2009ITS based Predictive Intelligent Battery Management System for plug-in Hybrid and Electric vehicles" Vehicle Power and Propulsion Conference, 2009. VPPC apos;09. IEEE Volume, 7-10710Sept. 2009 Page(s):138- 144

4. N. Brinkman, M. Wang, T. Weber, T. . Darlington, May, 2005Well-to-Wheels Analysis of Advanced Fuel/Vehicle Systems- A North American Study of Energy Use, Greenhouse Gas Emissions, and Criteria Pollutant Emissions

5. ASTME221302e1 Standard Specification for Telecommunications and Information Exchange Between Roadside and Vehicle Systems- 5 GHz Band Dedicated Short Range Communications (DSRC) Medium Access Control (MAC) and Physical Layer (PHY) Specifications

6. 802.11p – 2010 - IEEE Standard for Local and Metropolitan Area Networks- Specific requirements Part 11: Wireless LAN Medium Access Control (MAC) and Physical Layer (PHY) Specifications Amendment 6: Wireless Access in Vehicular Environments

7. IEEE 1609.x - IEEE Familty of Standards for Wireless Access in Vehicular Environments (WAVE)

8. SAE J2735 - SAE Standard for Dedicated Short Range Communications (DSRC) Message Set Dictionary

9. W. Fehr, 2011The Vehicle-to-Vehicle (2) Vand Vehicle-to-Infrastructure (V2I) Technology Test Bed- Test Bed 2.0: Available for Device and Application Development

10. Topcon2011Tripple Constellation Receiver

11. M. Boban, T. T. V. Vinhoza, 2011Modeling and Simulation of Vehicular Networks: towards Realistic and Efficient Models, Source: Mobile Ad-Hoc Networks: Applications, Book edited by: Xin Wang, 978-9-53307-416-0Publisher: InTech, Publishing date: January 2011

12. M. H. Lighthill, G. B. Whitham, 1955On kinematic waves II: A theory of traffic flow on long, crowded roadsProceedings of The Royal Society of London Ser. A 229317345

13. L. Bloomberg, J. Dale, 2000Comparison of VISSIM and CORSIM Traffic Simulation Models on a Congested NetworkTransportation Research Record 1727:52-60.

CHAPTER 9

Control of Hybrid Electrical Vehicles

Gheorghe Livinţ[1], Vasile Horga[1], Marcel Răţoi[1] and Mihai Albu[1]

[1] Gheorghe Asachi Technical University of Iaşi, Romania

1. INTRODUCTION

Developing cars is a major factor that has determined the increasing of the civilization degree and the continuous stimulation of the society progress. Currently, in Europe, one in five active people and in the US, one in four, directly work in the automotive industry (research, design, manufacture, maintenance) or in related domains (fuel, trade, traffic safety, roads, environmental protection). On our planet the number of the cars increases continuously and he nearly doubled in the last 10 years. With increasing number of cars entered in circulation every year, is held and increasing fuel consumption, increased environmental pollution due to emissions from internal combustion engines (ICE), used to their propulsion. Reducing oil consumption takes into account the limited availability of petroleum reserves and reducing emissions that affect the health of population in large urban agglomerations. The car needs a propulsion source to develop a maximum torque at zero speed. This can not be achieved with the classic ICE. For ICE power conversion efficiency is weak at low speeds and it has the highest values close to the rated speed. Pollution reduction can be achieved by using electric vehicles (EV), whose number is still significant. The idea of an electrical powered vehicle (EV) has been around for almost 200 years. The first electric vehicle was built by Thomas Davenport in 1834 [Westbrook, 2005]] But over time, the batteries used for energy storage could provide the amount of electricity needed to fully electric propulsion vehicles. Electric vehicles are powered by electric batteries which are charged at stations from sources supplied by electrical network with electricity produced in power plants. Currently, a lot of researches are focused on the possibility of using fuel cells for producing energy from hydrogen. EV with fuel cell can be a competitive alternative to the standard ICE that is used in today's cars. If performance is assessed overall thrust of the effort wheel and crude oil consumed for the two solutions: classic car with ICE and car with electric motor powered by electric batteries, the difference between their yields is not spectacular. In terms of exhaust emissions is the net advantage for electric vehicles. Pollutant emissions due to energy that is produced in power plants (plant property, located) are much easier to control than those produced by internal combustion engines of vehicles that are individual and scattered. Power plants are usually located outside urban areas, their emissions affects fewer people living in these cities. By using electric motors and controllers efficient, electric vehicles provide the means to achieve a clean and efficient urban transport system and a friendly environment. Electric vehicles are zero emission vehicles, called ZEV type vehicles (Zero-Emissions Vehicles).

Any vehicle that has more than one power source can be considered hybrid electric vehicle (HEV). But this name is used most often for a vehicle using for propulsion a combination of an electric drive motor and an ICE, which energy source is fossil fuel. The first patent for involving HEV technology was filed in 1905 by the american H. Piper. The change of focus to hybrid technology was done by almost all vehicle manufacturers. Many prototypes and a few mass produced vehicles are now available. For example, there were 23 hybrid electric presented at the North American International Auto Show (NAIAS) in 2000 [Wyczalek, 2000].

There are several configurations of electric and hybrid vehicles [Bayindir, 2011, Ehsani, 2005]: 1. electric vehicles equipped with electric batteries and/or supercapacitors called BEV (Battery Electric Vehicles), 2. hybrid electric vehicles which combine conventional propulsion based on ICE engine with petroleum fuel and electric propulsion with motor powered by batteries or supercapacitors called HEV (Hybrid Electric Vehicles), 3. electric vehicles equipped with fuel cells, called FCEV (Fuel Cell Electric Vehicles).

Concept of hybrid electric vehicle with ICE-electric motor aims to overcome the disadvantages of the pure electric vehicles, whose engines are powered by electric batteries: the limited duration of use (low autonomy) and time recharging for batteries.

2. HYBRID ELECTRIC VEHICLES

A hybrid electric vehicle is distinguee from a standard ICE driven by four different parts: a) a device to store a large amount of electrical energy, b) an electrical machine to convert electrical power into mechanical torque on the wheels, c) a modified ICE adapted to hybrid electric use, d) a transmission system between the two different propulsion techniques. Figure 1 shows the possible subsystems of a hybrid vehicle configuration [Chan, 2002], [Ehsani, 2005]

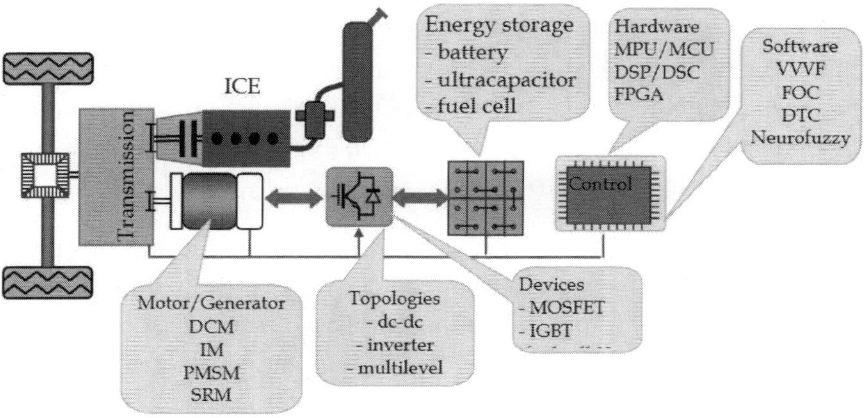

Figure 1. Main components of a hybrid electric vehicle

The devices used to store electrical energy could be batteries, hydrogen powered fuel cell or supercapacitors. Electric motors used on hybrid vehicles are [Husain 2003], [Fuhs, 2009]: DC motors, induction motors (IM), permanent magnet synchronous motors (PMSM) or switching reluctance motors (SRM). The HEV can use the electrical machine to behave as a generator and thereby produce electrical energy, which can be stored and used later. The ICE may be the same type as those on conventional vehicles, but it must be designed and optimized for hybrid vehicles. The transmission system between the ICE and the electrical machine is typically of series or parallel architecture. For power electronics are used MOSFET or IGBT transistors, and the command can be done with microprocessor, microcontroller or DSP using various techniques (VVVF - variable voltage and variable frequency, FOC –field oriented control, AC - adaptive control, NC – neural control or FC- fuzzy control).

Electric vehicles with two energy sources are also called hybrid vehicles. On hybrid-electric vehicles, in addition to the main battery, special batteries or capacitors, as a secondary energy source are used. These secondary energy sources are designed to provide power for short periods of peak operating conditions - for example, during the ascent of a slope or during acceleration. This is necessary because some batteries with the highest energy density have low power density. Since power density is required at least 150 [W/kg] for a good acceleration and slope climbing performance, a secondary source with high power density is essential. This power density is easily obtained from a lead-based battery and this is an auxiliary battery that is suitable for use with an aluminum-air battery in a hybrid-electric vehicle.

A combination of hybrid electric vehicle that is under development and of great interest, thanks to improvements in fuel cell, is the electric vehicle powered with fuel cell and an auxiliary battery. This battery can provide a high current necessary to start and can also serve as a load limiting device which allows the fuel cell to operate at low power first and then warm for a high power operation. This arrangement enhances the efficiency of the entire system and also allows the vehicle to use the recuperative braking.

Another class of hybrid electric vehicles, called hybrid electromechanical vehicles, use in addition to the main electric drive powered by batteries and a mechanical energy storage device such as a flywheel, or a hydraulic accumulator [Westbrook, 2005]. Hybrid electric vehicles represents a bridge between the present vehicle powered by ICEs and vehicles of the future characterized by a near-zero emissions, ULEV (Ultra-Low-Emission-Vehicle) or, in some cases even without pollution (ZEV-Zero-Emission Vehicle), as it is expected to be electrically propelled vehicles powered by fuel cells supplied with hydrogen.

It is very important to be reminded that without taking the technology steps and to improve the hybrid propulsion systems it is not possible to achieve higher level of the propulsion technology which uses fuel cells.

Currently a number of construction companies sell hybrid electric vehicles in series production: Toyota, Honda, Ford, General Motors. Many other companies have made prototypes of hybrid electric vehicles, the shift in mass production is only a matter of time that depends on the improvement of operating parameters and manufacturing cost reductions. Regarding the line of a hybrid electric vehicle powertrain, it is complex in terms of construction, operation and electronic control system than the most evolved similar vehicle equipped with conventional internal combustion engine.

Viewed from the standpoint of integration components, hybrid electric vehicle represents, compared with the vehicle solution powered ICE, an increase of complexity approximately 25%, while in terms of system control input hardware and software is at least double. These new elements make the price a such vehicle to be higher than that of a vehicle powered only by internal combustion engine. Thus, as shown in [Ehsani, 2005] the first vehicle (car) series hybrid electric Toyota designed by the company line is the most sophisticated and most advanced integrated powertrain control strategy of the firm ever made known to promote ideas of great ingenuity and technical complexity.

2.1. Architectures Of Hybrid Electric Drive Trains

The architecture of a hybrid vehicle is defined as the connection between the components of the energy flow routes and control ports. Hybrid electric vehicles were classified into two basic types: series and parallel.

But presently HEVs are classified into four kinds: series hybrid, parallel hybrid, series-parallel hybrid and complex. The primary power source (steady power source) is made up of fuel tank and ICE and battery-electric motor is taken as secondary source (dynamic power source).

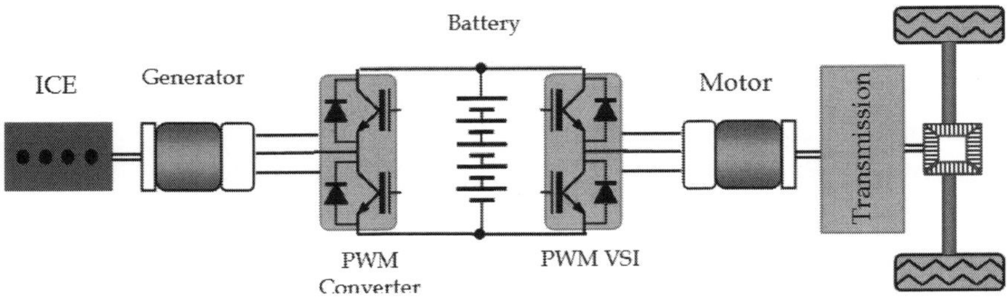

Figure 2. Series HEV

A series hybrid drive train uses two power sources which feeds a single powerplant (electric motor) that propels the vehicle. In Figure 2 is shown a series hybrid electric drive train where: fuel tank is an unidirectional energy source and the ICE coupled to an electric generator is a unidirectional energy converter. The electric generator is connected to an electric power bus through an electronic converter (rectifier). Electrochemical battery pack is the bidirectional energy source and is connected to the power bus by means of a power electronics converter (DC/DC converter). Also the electric power bus is connected to the controller of the electric traction motor. The traction motor can be controlled either as a motor (when propels the vehicle) or as generator (to vehicle braking). A battery charger can charge batteries with the energy provided by an electrical network.

The possible operating modes of series hybrid electric drive trains are [Ehsani, 2005]: 1. Pure electric: ICE is stopped and the vehicle is propelled only by batteries energy, 2. Pure engine mode: the vehicle is powered with energy provided by electric generator driven by engine. The batteries no provide and do not take energy from the drive train. 3. Hybrid mode: The traction power is drawn from both the engine-generator and the batteries. 4. Engine traction and battery charging mode: The ICE-generator provides the energy needed for the batteries charging and the propulsion vehicle. 5. Regenerative braking mode: the engine is turned off and the traction motor is operated as generator and the energy provided is used to charge the batteries. 6. Batteries charging mode: The engine –generator charges the batteries and the traction motor is not supplied. 7. Hybrid batteries charging mode: Both the engine-generator and the traction motor operate as generator to charge the batteries.

For series hybrid drive trains the following advantages can be mentioned: 1. Because the ICE is mechanical decoupled from the driven wheels and thus the ICE can operate solely within its maximum efficiency region, at a high-speed where and the emissions are reduced. 2. Electric motors have near-ideal torque-speed characteristics and multigear transmissions are not necessary. 3. The electrical transmission provides the mechanical decoupling and allows the using of a simple control strategy. The series hybrid electric drive trains have some disadvantages: 1. The mechanical energy of the ICE is first converted to electrical energy in the generator and the output energy of generator is converted into mechanical energy in the traction motor. The losses in the generator and the traction motor may be significant and these reduce the system efficiency. 2. The generator adds additional weight and cost. 3. The traction motor must be designed to meet maximum requirements, because only it is used for the vehicle propulsion.

Series hybrid electric vehicles applications include TEMSA – Avenue hybrid bus, Mercedes-Citaro bus and MAN-Lion's City Hybrid bus. Series hybrid configuration is mostly used in heavy vehicles, military vehicles and buses [Bayindir, 2011].

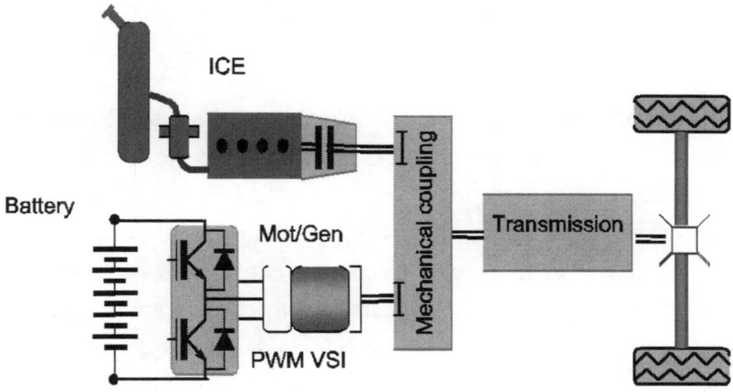

Figure 3. Parallel HEV

In the parallel configuration the power of the ICE and the electric motor are added into mechanical coupling, as shown in Figure 3 and operate the drive train by the mechanical transmission. There are different combination of the engine and electric motor power:

1. Torque-coupling parallel hybrid electric drive trains, 2. Speed-coupling parallel hybrid electric drive trains, 3. Torque-coupling and speed-coupling parallel hybrid electric drive trains. The torque coupling adds the torques of the engine *Tin1*, and the electric motor, *Tin2,* together or splits the engine torque into two parts: vehicle propelling and battery is charging. In this case the output torque, T_{out}, and speed, ω_{out}, can be described by

$$T_{out} = k_1 T_{in1} + k_2 T_{in2} \quad , \quad \omega_{out} = \frac{\omega_1}{k_1} = \frac{\omega_2}{k_2} \tag{1}$$

where *k1* and *k2* are the constants determined by the parameters of torque coupling.

The speed coupling adds the speeds of the engine, ω_{in1} , and the electric motor, ω_{in2}, together by coupling their speeds. The output speed ω_{out} and torque, T_{out} , can be described by:

$$\omega_{out} = k_1 \omega_{in1} + k_2 \omega_{int2} \quad , \quad T_{out} = \frac{T_{in1}}{k_1} + \frac{T_{in2}}{k_2} \tag{2}$$

where *k1* and *k2* are the constants determined by the parameters of speed coupling.

In parallel hybrid electric vehicle various control strategies can be used. In the most common strategies, ICE is used in on mode and operates at almost constant power output at its area of maximum efficiency. If the power requested from drive train is higher than the output power of ICE, the electric motor is turned on, ICE and electric motor supply power to the drive train. If the power requested from drive train is less than the output power of the ICE, the remaining power is used for charging the batteries. In this configuration, regenerative braking power on a down slope driving is used to charge the batteries. Examples of the parallel hybrid electric vehicles are [Fuhs, 2009]: Insight model introduced by Honda, Ford Escape Hybrid SUV and Lexus Hybrid SUV.

3. ENERGY STORAGE FOR HEV

For hybrid electric vehicles three electromechanical devices are important: batteries supercapacitors and fuel cels (FC). The batteries provide storage of energy that is essential for regenerative braking. Also the batteries are a source of energy that is necessary for electric-only propulsion. Supercapacitors, which have the similar capability, are high-power and low-stored-energy devices. Supercapacitors are used to improve high power peak for short duration. Batteries are heavy and repeated deep discharge adversely affects life. Batteries have much better performance as a provider of peak power for hybrid electric vehicle and/or fuel cell vehicle.

The fuel cell (FC) system has a fuel tank and stores energy in the form of hydrogen. An FC can be used for electric-only propulsion and it cannot be used for regenerative braking.

The batteries can be classified in primary batteries, which cannot be recharged, and secondary batteries, which have reversible reactions and can be charged and discharged. The basic unit of a battery is the voltaic cell. Voltaic cells have the following components parts: the electrodes: anode and cathode, and the electrolyte. Batteries are composed of collections of cells. Each cell has a voltage which depends of the electrochemical potential of the chemicals. Lead acid has a cell potential of 2 V, and Nickel metal hydride (NiMH) has cell potential of 1.2 V. The cells are connected in series for increasing the voltage. Batteries are described by four quantitative features: current, voltage, energy and power. Capacity (C) is the charge that can be taken from the battery under certain conditions, so is the amount of electricity that a battery can provide in continuous discharging. A constant current battery capacity is a function of current values to which they discharge. Battery capacity is expressed in the [Ah] and is given by $C=\int I_d dt$ where I_d is the discharge current and t is the time of discharge. If the discharge is made at constant current, $C=I_d t$. Cell capacity, C, is determined partially by the mass of available reactants.

3.1. Batteries Characteristics

The main batteries characteristics are: specific energy, specific power, self discharge, life, state of charge (SOC).

Specific energy is the ratio between the energy provided of battery and its weight and is expressed in [Wh/kg]. It should be noted that the specific energy storage of gasoline is about 11000 [Wh/kg] versus 90 [Wh/kg] representing a peak through the current batteries It noted that lead-based batteries have the lowest energy density as lithium batteries have the highest energy densities. **Specific power**, expressed in [W/kg] or [W/l] represents the ratio between the battery power and its weight or between battery power and its volume. Self discharge consists in the battery power loss due to internal reactions during storage/non-use them, when the external circuit is open (missing load). Self discharge rate depends on battery type, temperature and battery age. Life (number of cycles) is expressed in number of cycles, i.e. the number of complete charge and discharge (until the final voltage permissible or until capacity drops below 80% of initial capacity) that the battery can handle. State of charge - SOC is the ratio of the electric charge *Q* than can be delivered by the battery at current *I*, to the nominal battery capacity *Q0*

$$SOC = \frac{Q(t)}{Q_0} = \frac{Q_0 - It}{Q_o} \tag{3}$$

Knowing the energy remaining from the initial energy, we get an indication of the time as the battery will continue to work before its charging. SOC decreases as the battery supplies a load and increase SOC when the battery is charging. Note that the SOC does not refer to the maximum capacity that the battery was last charged, but to the initial battery capacityThe degree of discharge (Depth of Discharge - DOD) on each cycle represents the percentage of energy discharged / removed from the battery to each discharge.

3.2. Batteries Models Used In Hybrid Electric Vehicles

The simplest electrical battery-model is presented in Figure 4.a. It consist of an ideal voltage source *VOC*, a constant equivalent internal resistance *Ri* and the terminal voltage represented by *VO* [Livint et. all. 2007]. However, the internal resistance is different under discharge and charge conditions. This model does not capture the internal dynamics of the battery, in particular the effect of the diffusion of the electrolytic chemicals between the battery plates. The circuit can be modified as shown in Figure 4.b. for the account the different resistance values under charge and discharge conditions, *Rc* and *Rd*. The diodes shown in Figure 4.b have no physical significance in the battery and are included only for modeling purposes.

In order to model the diffusion of the electrolytic through the battery and its resultant effect of causing transient currents in the battery, a capacitor *C* is added to the model. A dynamical model of the lead-acid battery, presented in Figure 4.c, shows a simplified equivalent circuit proposed in [Livint et al., 2006].

The cell terminal voltage is represented by *VO*, and *Ri* is a lumped resistance due to cell interconnections. A double layer capacitance *Cs* (surface-capacitor) is shown in parallel with the charge transfer polarization represented by *Rt*. This double layer capacitor is the results of charge separation at the electrolyte/electrode interface.

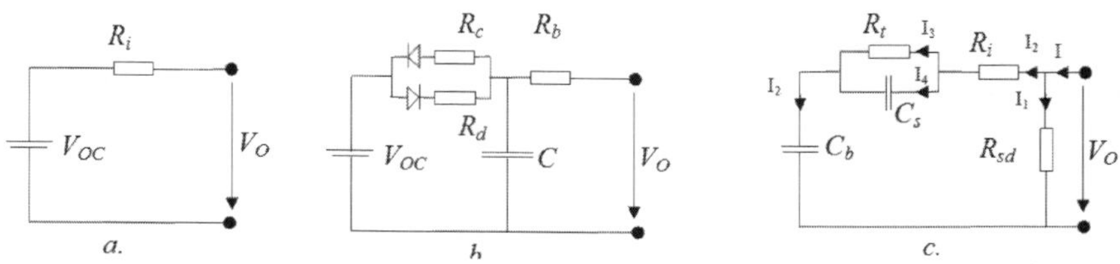

Figure 4. a. Simple battery model; b. Battery model with the different charging and discharging resistance values and polarization capacitance; c. Simplified equivalent circuit of a lead acid battery

The bulk capacitor *Cb* models the cell's open circuit voltage, and *Rsd* is included to represent the self-discharge of the cell.

Voltages and currents describing the characteristics of the model shown in Fig.4.c are given by equations

$$
\begin{aligned}
& V_O = R_i I_2 + V_{Cs} + V_{Cb} = R_{sd} I_1 \\
& V_{Cs} = \frac{1}{C_s}\int I_4 dt = R_t I_3 \\
& V_{Cb} = \frac{1}{C_b}\int I_2 dt \\
& I = I_1 + I_2 ; I_2 = I_3 + I_4
\end{aligned}
\tag{4}
$$

where *VCb* and *VCs* denote the voltages across the bulk- and surface-capacitors, respectively

Taking as state variables the voltages *VO*, *VCs*, *VCb* and assuming that $dI_2/dt \approx 0$ (the rate of change of terminal current per sampling interval, when implemented digitally, is negligible) the complete state variable description (the state model of battery) is obtained:

$$
\begin{bmatrix} \dot{V}_O \\ \dot{V}_{Cs} \\ \dot{V}_{Cb} \end{bmatrix} = \begin{bmatrix} -\frac{1}{R_{sd}}\left(\frac{1}{C_s}+\frac{1}{C_b}\right) & -\frac{1}{R_t C_s} & 0 \\ -\frac{1}{R_{sd} C_s} & -\frac{1}{R_t C_s} & 0 \\ -\frac{1}{R_{sd} C_b} & 0 & 0 \end{bmatrix} \begin{bmatrix} V_O \\ V_{Cs} \\ V_{Cb} \end{bmatrix} = \begin{bmatrix} \frac{1}{C_s}+\frac{1}{C_b} \\ \frac{1}{C_s} \\ \frac{1}{C_b} \end{bmatrix} I
\tag{5}
$$

$$
y = V_O = [1 \;\; 0 \;\; 0][V_O \;\; V_{Cs} \;\; V_{Cs}]^T
$$

The control strategies for hybrid and electric vehicles are based on SOC knowledge. The batteries behavior can be modeled with different electrical circuit structures and different linear or nonlinear mathematical models. The different estimation methods can be used for the batteries parameters determination. The method based on continue time model and LIF algorithm assures a good accuracy for parameters estimation. The estimated parameters can be used for the battery SOC determination.

4. ELECTRIC MOTORS USED FOR HYBRID ELECTRIC VEHICLES PROPULSION

4.1. Motor Characteristics Versus Electric Traction Selection

The electric motors can operate in two modes: a) as motor which convert electrical energy taken from a source (electric generator, battery, fuel cell) into mechanical energy used to propel the vehicle, b) as generator which convert the mechanical energy taken from a motor (ICE, the wheels during vehicle braking, etc..) in electrical energy used for charging the battery. The motors are the only propulsion system for electric vehicles. Hybrid electric vehicles have two propulsion systems: ICE and electric motor, which can be used in different configurations: serial, parallel, mixed. Compared with ICE electric motors has some important advantages: they produce large amounts couples at low speeds, the instantaneous power values exceed 2-3 times the rated ICE, torque values are easily reproducible, adjustment speed limits are higher. With these characteristics ensure good dynamic performance: large accelerations and small time both at startup and braking.

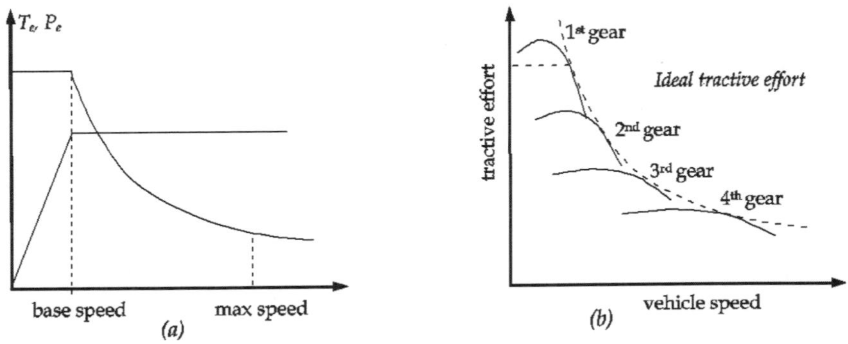

Figure 5. a. Characteristics of traction motors ; b. Tractive effort characteristics of an ICE vehicle

Figure 5.a illustrates the standard characteristics of an electric motor used in EVs or HEVs.

Indeed, in the constant-torque region, the electric motor exerts a constant torque (rated torque) over the entire speed range until the rated speed is reached. Beyond the rated speed of the motor, the torque will decrease proportionally with speed, resulting in a constant power (rated power) output. The constant-power region eventually degrades at high speeds, in which the torque decreases proportionally with the square of the speed. This characteristic corresponds to the profile of the tractive effort versus the speed on the driven wheels [Figure 5. b.]. This profile is derived from the characteristics of the power source and the transmission. Basically, for a power source with a given power rating, the profile of the tractive effort versus the speed should be a constant.

The power of the electric motor on a parallel type hybrid vehicle decisively influences the dynamic performance and fuel consumption. The ratio of the maximum power the electric motor, *PEM*, and ICE power *PICE*is characterized by hybridization factor which is defined by the relation *HF*

$$HF = \frac{P_{EM}}{P_{EM} + P_{ICE}} = \frac{P_{EM}}{P_{HEV}} \tag{6}$$

where *PHEV* is the maximum total traction power for vehicle propulsion. It is demonstrated that it reduces fuel consumption and increase the dynamic performance of a hybridization factor optimal point more than one critic (*HF*=0.3÷0.5) above the optimal point increase in ICE power hybrid electric vehicle does not improve performance.

The major requirements of HEVs electric propulsion, as mentioned in literature, are summarized as follows [Chan 2005], [Husain 2003], [Ehsani 2005]:

1. a high instant power and a high power density;
2. a high torque at low speeds for starting and climbing, as well as a high power at high speed for cruising;
3. a very wide speed range, including constant-torque and constant-power regions;
4. a fast torque response;

5. a high efficiency over the wide speed and torque ranges;
6. a high efficiency for regenerative braking;
7. a high reliability and robustness for various vehicle operating conditions; and
8. a reasonable cost.

Moreover, in the event of a faulty operation, the electric propulsion should be fault tolerant. Finally, from an industrial point of view, an additional selection criterion is the market acceptance degree of each motor type, which is closely associated with the comparative availability and cost of its associated power converter technology [Emadi 2005].

4.2. Induction Motors Used In Hybrid Electric Vehicles

4.2.1. Steady State Operation of Induction Motor

Induction motor is the most widely used ac motor in the industry. An induction motor like any other rotating machine consists of a stator (the fixed part) and a rotor (the moving part) separated by air gap. The stator contains electrical windings housed in axial slots. Each phase on the stator has distributed winding, consisting of several coils distributed in a number of slots. The distributed winding results in magnetomotive forces (MMF) due to the current in the winding with a stepped waveform similar to a sine wave. In three-phase machine the three windings have spatial displacement of 120 degrees between them. When balanced three phase currents are applied to these windings, the resultant MMF in the air gap has constant magnitude and rotates at an angular speed of $\omega_s = 2\pi fs$ electrical radians per second. Here fs is the frequency of the supply current. The actual speed of rotation of magnetic field depends on the number of poles in the motor. This speed is known as synchronous speed Ω_s of the motor and is given by

$$\Omega_s = \frac{\omega_s}{p} = \frac{2\pi f_s}{p} = \frac{2\pi n_s}{60}; \; n_s = \frac{60 f_s}{p} \tag{7}$$

where p is number of pole pairs, *ns* [rpm], is the synchronous speed of rotating field.

If the rotor of an induction motor has a winding similar to the stator it is known as wound rotor machine. These windings are connected to slip rings mounted on the rotor. There are stationary brushes touching the slip rings through which external electrical connected. The wound rotor machines are used with external resistances connected to their rotor circuit at the time of starting to get higher starting torque. After the motor is started the slip rings are short circuited. Another type of rotor construction is known as squirrel cage type rotor. In this construction the rotor slots have bars of copper or aluminium shorted together at each end of rotor by end rings. In normal running there is no difference between a cage type or wound rotor machine as for as there electrical characteristics are concerned.

When the stator is energized from a three phase supply a rotating magnetic field is produced in the air gap. The magnetic flux from this field induces voltages in both the stator and rotor windings. The electromagnetic torque resulting from the interaction of the currents in the rotor circuit (since it is shorted) and the air gap flux, results in rotation of rotor. Since electromotive force in the rotor can be induced only when there is a relative motion between air gap field and rotor, the rotor rotates in the same direction as the magnetic field, but it will not run at synchronous speed. An induction motor therefore always runs at a speed less than synchronous speed. The difference between rotor speed and synchronous speed is known as slip. The slip s is given by

$$s = \frac{\Omega_s - \Omega}{\Omega_s} = \frac{\Omega_{sl}}{\Omega_s} = 1 - \frac{\Omega}{\Omega_c}; or: s = \frac{n_s - n}{n_s} = \frac{n_{sl}}{n_s} = 1 - \frac{n}{n_s} \tag{8}$$

where: n [rpm] it the speed of the rotor.

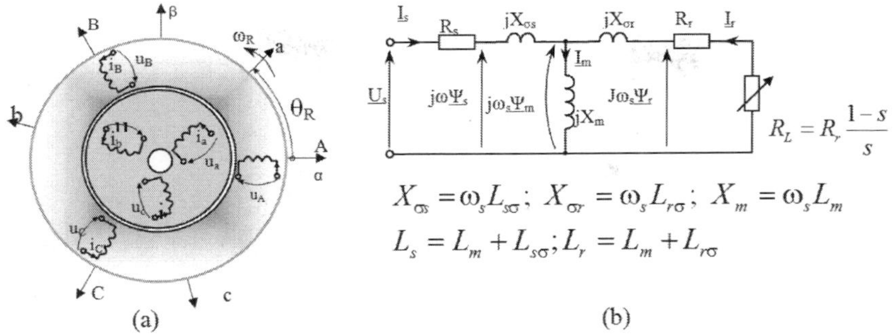

Figure 6. Cross section of an induction motor (a); Equivalent circuit of an IM (b)

The steady state characteristics of induction machines can be derived from its equivalent circuit. In order to develop a per phase equivalent circuit of a three-phase machine, a wound rotor motor as shown in Figure 6.a. is considered here. In case of a squirrel cage motor, the rotor circuit can be replaced by an equivalent three-phase winding. When three-phase balanced voltages are applied to the stator, the currents flow in them. The equivalent circuit, therefore is identical to that of a transformer, and is shown in Figure 6. b. Here *Rs* is the stator winding resistance, *Ls* is self inductance of stator, *Lr*is self inductance of rotor winding referred to stator, *Rr* is rotor resistance referred to stator, *Lm* is magnetizing inductance and *s* is the slip. The parameters of the equivalent circuit are the stator and rotor leakage reactances $X_{\sigma s}$ and $X_{\sigma r}$, magnetizing reactance *Xm*, and the equivalent resistance $R_L = \frac{1-s}{s} R_r$ which depends on the slip *s*.

The ohmic losses on this "virtual" resistance, *RL,* represent the output mechanical power, *Pmec*, transferred to the load. Thus the electromagnetic torque, *Te*, is given as

$$T_e = \frac{P_{mec}}{\Omega} = \frac{3p}{(1-s)\omega_s} \frac{(1-s)R_r}{s} I_r^2 = \frac{3p}{\omega_s} \frac{R_r}{s} I_r^2 \tag{9}$$

If statoric leakage reactance is neglected it results

$$T_e = 3p\left(\frac{U_s}{\omega_s}\right)^2 \frac{R_r s}{R_r^2 + (\omega_{sl} L_{r\sigma})^2} \tag{10}$$

For applications where high degree of accuracy in speed control is not required simple methods based on steady state equivalent circuit have been employed. Since the speed of an induction motor, *n*, in revolutions per minute is given by

$$n = \frac{60 f_s}{p}(1-s) \tag{11}$$

Thus the speed of the motor can be changed by controlling the frequency, or number of poles or the slip. Since, number of poles of a motor is fixed at the time of construction, special motors are required with provision of pole changing windings.

4.2.2. The Dynamic Model of The Induction Motor

The dynamic model of ac machine can be developed [Ehsani, 2005], [Husain, 2003], using the concept of "space vectors". Space vectors of three-phase variables, such as the voltage, current, or flux, are very convenient for the analysis and control of ac motors and power converter. A three-phase system defined by *yA(t), yB(t), and yC(t)* can be represented uniquely by a rotating vector $\underline{y}(t)$ in the complex plane.

$$\underline{y}(t) = \frac{2}{3}(y_A(t) + \underline{a} \cdot y_B(t) + \underline{a}^2 \cdot y_C(t)) = y_D(t) + j \cdot y_Q(t) \tag{12}$$

Where $a = e^{j(2\pi/3)}$

Under simplifying assumptions (symmetrical windings with sinusoidal distribution, negligible cross-section of the conductors, ideal magnetic circuit) the induction squirrel cage machine may be described in an arbitrary synchronous reference frame, at ωg speed, by the following complex space vector equations [Livint et all 2006]:

$$\begin{cases} \underline{u}_{sg} = R_s \cdot \underline{i}_{sg} + \dfrac{d\underline{\psi}_{sg}}{dt} + j\omega_g \underline{\psi}_{sg}; \quad 0 = R_r \cdot \underline{i}_{rg} + \dfrac{d\underline{\psi}_{rg}}{dt} + j(\omega_g - \omega_r)\underline{\psi}_{rg} \\ \underline{\psi}_{sg} = L_s \underline{i}_{sg} + L_m \underline{i}_{rg}; \quad \underline{\psi}_{rg} = L_m \underline{i}_{sg} + L_r \underline{i}_{rg} \\ t_e = -\dfrac{3}{2} p L_m \underline{i}_{sg} \times \underline{i}_{rg}; \quad J\dfrac{d\Omega}{dt} = t_e - D\Omega - t_l \end{cases} \quad (13)$$

where: $\underline{x}_{sg} = \underline{x}_s e^{-j\theta_g}$; $\underline{x}_{rg} = \underline{x}_r e^{-j(\theta_g - \theta_r)}$; $\omega_g = \dfrac{d\theta_g}{dt}$ - speed of the arbitrary reference frame,

$\omega_r = p\Omega = \dfrac{d\theta_r}{dt}$ - speed of the rotor reference frame.

In order to achieve the motor model in stator reference frame on impose ωg=0, in equations (13).

4.3. Power Converters

Power converters play a vital role in Hybrid Electric Vehicle (HEV) systems. Typical HEV drive train consists of a battery, power converter, and a traction motor to drive the vehicle. The power converter could be just a traditional inverter or a dc-dc converter plus an inverter. The latter configuration provides more flexibility and improves the system performance. The dc-dc converter in this system interfaces the battery and the inverter dc bus, and usually is a variable voltage converter so that the inverter can always operate at its optimum operating point. In most commercially available systems, traditional boost converters are used. A power converter architecture is presented in Figure 7.

Voltage source inverters (VSI) are used in hybrid vehicles to control the electric motors and generators. The switches are usually IGBTs for high-voltage high power hybrid configurations, or MOSFETs for low-voltage designs. The output of VSI is controlled by means of a pulse-witth-moduated (PWM) signal to produce sinusoidal waweform. Certain harmonics exist in such a switching scheme. High switching frequency is used to move the armonics away from the fundamental frequency.

A three-phase machine being feed from a VSI receives the symmetrical rectangular three-phase voltages shown in Figure 8.a. Inserting these phase voltage in the space vector definition of stator voltage $\underline{u}_S(t) = \frac{2}{3}\left(u_{SA}(t) + \underline{a}u_{SB}(t) + \underline{a}^2 u_{SC}(t)\right)$, yields the typical set of six active switching state vectors *U1* ... *U6* and two zero vectors *U0* and *U7* as shown in Figure 8.b.

$$\underline{u}_s = \begin{cases} \frac{2}{3} U_{dc} e^{j \cdot k \cdot 2\pi/3} & k = 1,..,6 \\ 0 & k = 0,7 \end{cases} \tag{14}$$

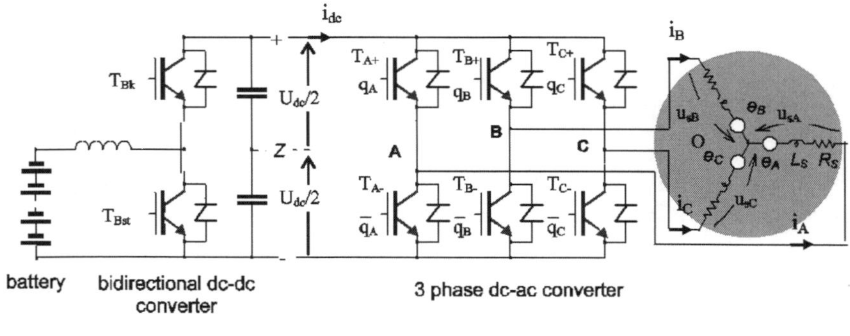

Figure 7. Power converter architecture

5. CONTROL STRATEGIES

A number of control strategies can be used in a drive train for vehicles with different mission requirements. The control objectives of the hybrid electric vehicles are [Ehsani, 2005]: 1) to meet the power demand of the driver, 2) to operate each component of the vehicle with optimal efficiency, 3) to recover braking energy as much as possible, 4) to maintain the state-of-charge (SOC) of the battery in a preset window.

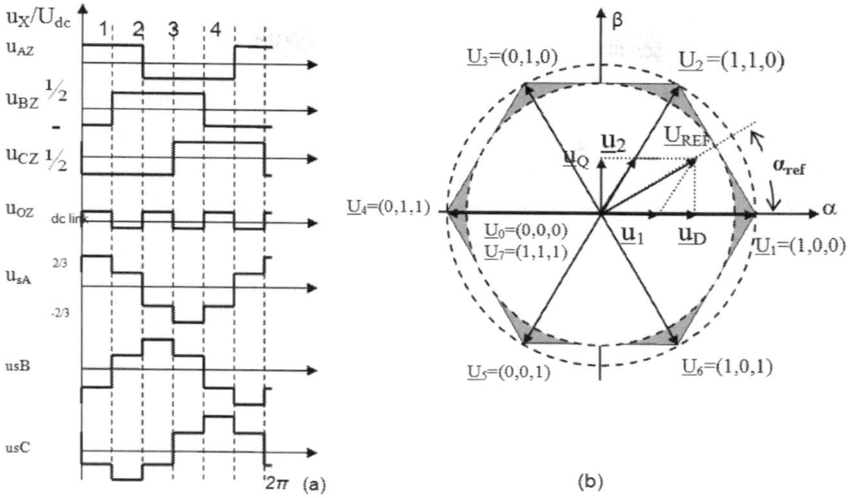

Figure 8. a. Switched three-phase waveforms ; b. Switching state vectors

The induction motor drive on EV and HEV is supplied by a DC source (battery, fuel cell,) which has a constant terminal voltage, and a DC/AC inverter that provide a variable frequency and variable voltage. The DC/AC inverter is constituted by power electronic switches and power diodes.

As control strategies PWM control is used for DC motor, FOC (field-oriented control) and DTC (direct torque control) are used for induction motors. The control algorithms used are the classical control PID, but and the modern high-performance control techniques: adaptive control, fuzzy control, neuro network control [Seref 2010], [Ehsani 2005], [Livint et all 2008a, 2008c].

5.1. Structures For Speed Scalar Control Of Induction Motor

5.1.1. Voltage And Frequency (Volts/Hz) Control

Equation (11) indicates that the speed of an induction motor can be controlled by varying the supply frequency f_s. PWM inverters are available that can easily provide variable frequency supply with good quality output wave shape. The open loop volts/Hz control is therefore quite popular method of speed control for induction motor drives where high accuracy in control is not required. The frequency control also requires proportional control in applied voltage, because then the stator flux $\Psi s = Us/\omega s$(neglecting the resistance drop) remains constant. Otherwise, if frequency alone is controlled, then the flux will change. When frequency is increased, the flux will decrease, and the torque developed by the motor will decrease as shown in Figure 9.a. When frequency is decreased, the flux will increase and may lead to the saturation of magnetic circuit. Since in PWM inverters the voltage and frequency can be controlled independently, these drives are fed from a PWM inverter.

The control scheme is simple as shown in Figure 9.b with motor being supplied by three-phase supply dc-link and PWM inverter.

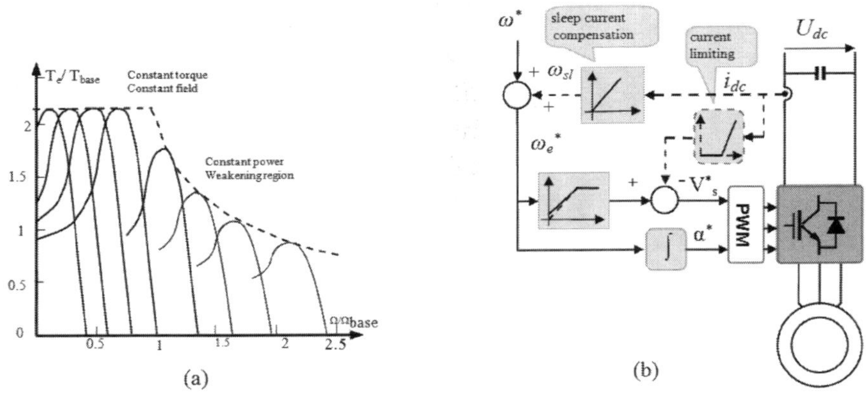

Figure 9. a. Torque-speed characeristics under V/f control; b. VSI induction motor drive V/f controlled

The drive does not require any feedback and is used in low performance applications where precise speed control is not required. Depending on the desired speed the frequency command is applied to the inverter, and phase voltage command is directly generated from the frequency command by a gain factor, and input dc voltage of inverter is controlled.

The speed of the motor is not precisely controlled by this method as the frequency control only controls the synchronous speed [Emadi, 2005], [Livint et al. 2006] There will be a small variation in speed of the motor under load conditions. This variation is not much when the speed is high. When working at low speeds, the frequency is low, and if the voltage is also reduced then the performance of the motor are deteriorated due to large value of stator resistance drop. For low speed operation the relationship between voltage and frequency is given by

$$U_s = U_0 + kf_s \tag{15}$$

where *U0* is the voltage drop in the stator resistance.

5.2 Structures For Speed Vector Control Of Induction Motor

In order to obtain high performance, and fast dynamic response in induction motors, it is important to develop appropriate control schemes. In separately excited dc machine, fast transient response is obtained by maintaining the flux constant, and controlling the torque by controlling the armature current.

The vector control or *field oriented control* (FOC) of ac machines makes it possible to control ac motor in a manner similar to the control of a separately excited dc motor. In ac machines also, the torque is produced by the interaction of current and flux. But in induction motor the power is fed to the stator only, the current responsible for producing flux, and the current responsible for producing torque are not easily separable. The basic principle of vector control is to separate the components of stator current responsible for production of flux, and the torque. The vector control in ac machines is obtained by controlling the magnitude, frequency, and phase of stator current, by inverter control. Since, the control of the motor is obtained by controlling both magnitude and phase angle of the current, this method of control is given the name vector control.

In order to achieve independent control of flux and torque in induction machines, the stator (or rotor) flux linkages phasor is maintained constant in its magnitude and its phase is stationary with respect to current phasor.

The vector control structure can be classified in: 1. direct control structure, when the oriented flux position is determined with the flux sensors and 2. indirect control structure, then the oriented flux position is estimated using the measured rotor speed.

For indirect vector control, the induction machine will be represented in the synchronously rotating reference frame. For indirect vector control the control equations can be derived with the help of d-q model of the motor in synchronous reference frame as given in 13.

The block diagram of the rotor flux oriented control a VSI induction machine drive is presented inFigure 10.

Generally, a closed loop vector control scheme results in a complex control structure as it consists of the following components: 1. PID controller for motor flux and toque, 2. Current and/or voltage decoupling network, 3. Complex coordinate transformation, 4. Two axis to three axis transformation, 5. Voltage or current modulator, 6. Flux and torque estimator, 7. PID speed controller

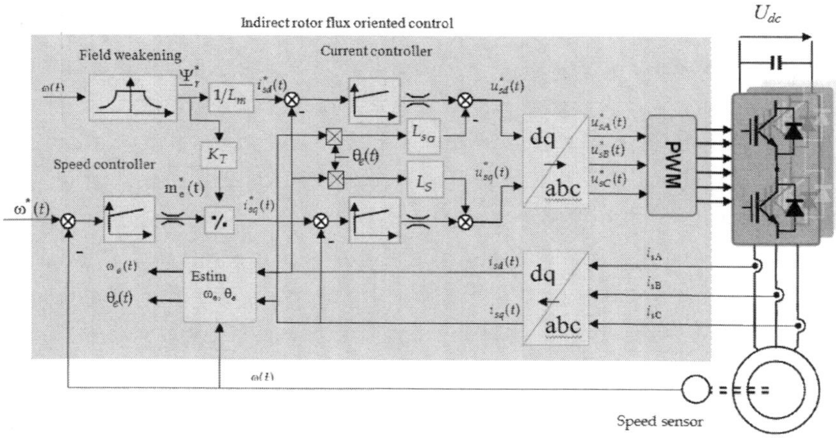

Figure 10. Block diagram of the rotor flux oriented control of a VSI induction machine drive

6. EXPERIMENTAL MODEL OF HYBRID ELECTRIC VEHICLE

The structure of the experimental model of the hybrid vehicle is presented in Figure 11. The model includes the two power propulsion (ICE, and the electric motor/generator M/G) with allow the energetically optimization by implementing the real time control algorithms. The model has no wheels and the longitudinal characteristics emulation is realized with a corresponding load system. The ICE is a diesel F8Q of 1.9l capacity and 64[HP]. The electronic unit control (ECU) is a Lucas DCN R04080012J-80759M. The coupling with the motor/generator system is assured by a clutch, a gearbox and a belt transmission.

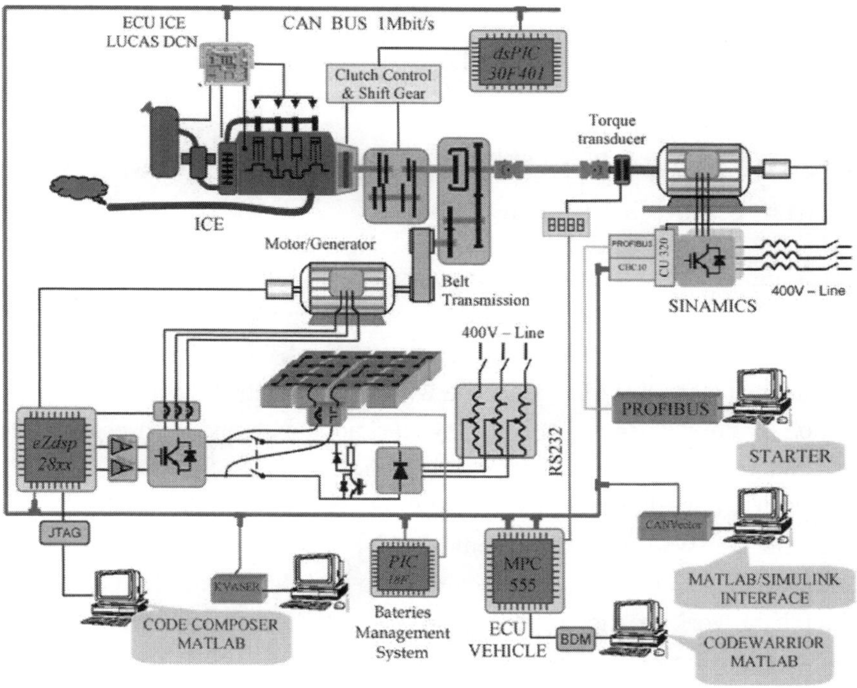

Figure 11. The structure of the experimental model of the hybrid electric vehicle

The electric machine is a squirrel cage asynchronous machine (15kW, 380V, 30.5A, 50Hz, 2940 rpm) supplied by a PWM inverter implemented with IGBT modules (SKM200GB122D). The motor is supplied by 26 batteries (12V/45Ah). The hardware structure of the motor/generator system is presented in Figure 12. The hardware resources assured by the control system eZdsp 2808 permit the implementation of the local dynamic control algorithms and for a CAN communication network, necessary for the distributed control used on the hybrid electric vehicle, [Livint et all 2008, 2010]

With the peripheral elements (8 ePWM channels, 2x8 AD channels with a resolution of 12 bits, incremental transducer interface eQEP) and the specific peripheral for the communication assure the necessary resources for the power converters command and for the signal acquisition in system. For the command and state signal conditioning it was designed and realized an interface module.

6.1. The Emulation Of The Longitudinal Dynamics Characteristics Of The Vehicle

The longitudinal dynamics characteristics of the vehicle are emulated with an electric machine with torque control, Figure 13. As a mechanical load emulator, the electric machine operates both in motor and generator regimes. An asynchronous machine with vector control technique assures a good dynamic for torque. This asynchronous machine with parameters (15KW, 28.5A, 400V, 1460rpm) is supplied by a SINAMICS S120 converter from Siemens which contains a rectifier PWM, a voltage dc link and a PWM inverter [Siemens 2007]. This converter assures a sinusoidal current at the network interface and the possibility to recover into the network the electric energy given by the electric machine when it operates in generator regime.

The main objective is to emulate the static, dynamics and operating characteristics of the drive line. The power demand for the vehicle driving at a constant speed and on a flat road [Ehsani, 2005], can be expressed as

$$P_e = \frac{v}{1000\eta_{t,e}}(m_v g f_r + \frac{1}{2}\rho_a C_D A_f v^2 + m_v g i) \ [kW] \tag{16}$$

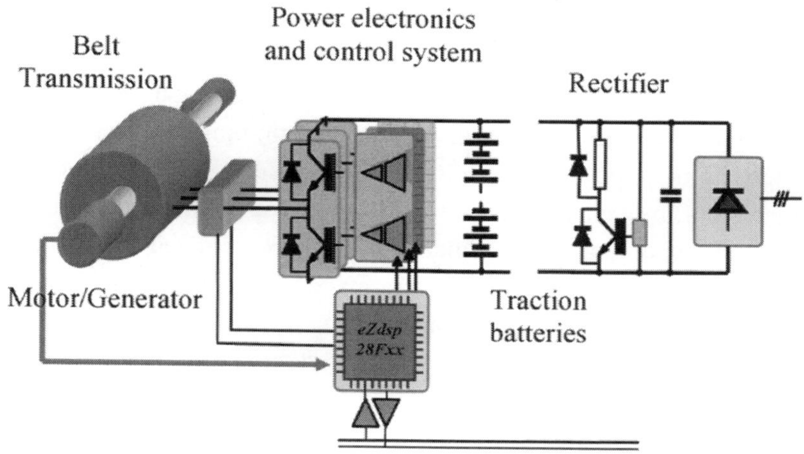

Figure 12. Electric motor/generator system

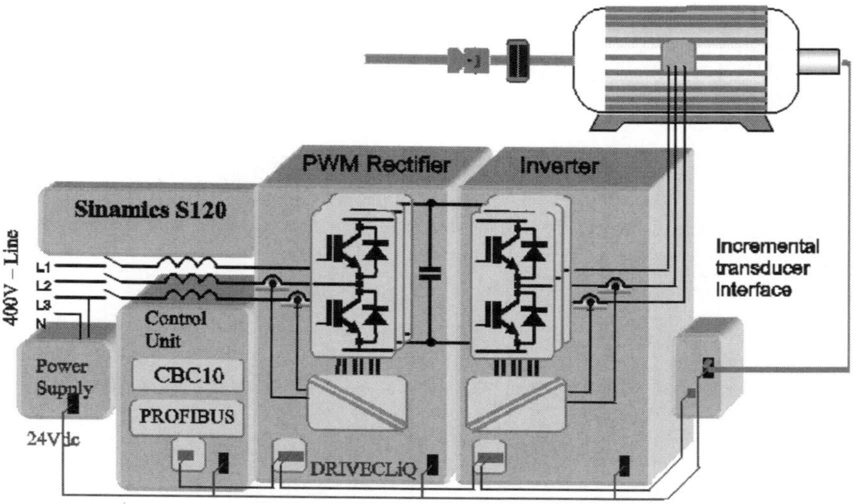

Figure 13. Emulation system of the longitudinal dynamics characteristics of the vehicle

6.2. The Distributed System Of The Real-Time Control Of The Hybrid Electric Vehicle Model

The coordinated control of the sub-systems of the parallel hybrid vehicle can be realized with a hierarchical structure, [Livint et. al, 2006, 2008]. Its main element is the Electronic Control Unit vehicle of the vehicle (ECU vehicle) which supervises and coordinates the whole systems.

It has to monitor permanently the driver demands, the motion conditions and the state of the sub-systems in order to estimate the optimum topology of the whole system and to assure minimum fuel consumption at high running performances. The main system must to assure the maneuverability demanded by the driver in any running conditions. These supervising and coordinating tasks are realized by a control structure that includes both state automata elements and dynamic control elements corresponding to each state. The dynamic control of each sub-system is realized by every local control system. The dynamic control is integrated at the level of the coordinating system only when it is necessary a smooth transition between states or for a dynamic change into a state with more than a sub-system (starting engine with the electric machine).

The optimization of the performances objectives is realized logically by the state automata. The optimum operating state is determined by the coordinating and supervising system based on the analysis of the centralized data.

The state machine design is achieved in three stages:

- the identification of the all possible operating modes of the vehicle,
- the evaluation of the all possible transitions between the operating modes,
- the arbitration of the priorities between the concurrent transitions.

For the first stage it is realized a list with the possible operating modes for each sub-system. For example, for the engine the possible operating modes are running engine and stop engine.

After the identification of the all possible operating modes of each subsystem, it is generated a set of all the possible combinations of the operating modes for the vehicle.

Due to the complexity of the real time control for a parallel hybrid electric vehicle it is necessary to integrate all the elements in a high speed CAN communication network (1Mbps) to assure the distributed control of all resources [CANopen, 2004], [Chacko, 2005]. The experimental model uses a CANopen network with four slave nodes and one master node. The master node is implemented on phyCORE-mpc555 system and assures the network management and supervises the nodes control connected by NMT services, the nodes operating states, the emergency messages analysis and the modifications appeared into the communication network. The first CANopen slave node, at an inferior level, is dedicated for the motor/generator system and includes the speed control loop for the vehicle electrical propulsion.The second slave node is used to take over the torque data given on the RS-232 serial line by the DTR torque transducer and to convert the data for the proper utilization on the CANopen network.

The third slave node of the CANopen network is used for the emulation system for the longitudinal dynamics characteristics of the vehicle, implemented with the asynchronous motor and the SINAMICS S120 converter.

The fourth slave node of the CANopen network is the system of automatic gear shift, which involves control of clutch and gear. Control is achieved with a numerical dsPIC-30F4011.

The CAN protocol utilizes versatile message identifiers that can be mapped to specific control information categories. With predefined priority of the communication message, non-destructive bit-wise arbitration with error detection signaling, the CAN protocol supports distributed real-time control in vehicles applications with a very high level of security.

The content of a message is named by an *identifier*. The *identifier* describes the meaning of the date, but not indicates the destination of the message. All nodes in the network are able to decide by *message filtering* whether the data is to be accepted. If two or more nodes attempt to transmit at the same time, a non-destructive arbitration technique guarantees the messages transmission in order of priority and that no messages are lost.

It is guaranteed that a message is simultaneously accepted by all nodes of a CAN network. When a receiver detects an error in the last bit that cares about it will send an error frame and the transmitter will retransmit the message.

The CAN network provides standardized communication objects for real-time data (PDO – Process Data Objects), configuration data (SDO – Service Data objects), and special functions (Emergency message), network management data (NMT message, Error control).

Service Data Object (SDO) supports the mandatory OD (Object Dictionary) entries, slave support for the next slave services: Reset_Node, Enter_Preoperational_State, Start_Remote_Node, Stop_Remote_Node, Reset_Communication, COB (Communication Data Object).

For the software design it was in attention the modularity and a scalar structure of the final product that can be easy configured for the automation necessities of the communication node. For this the CANopen stack was structured in two modules [Livint et all, 2008, 2009]:

- Module I, dependent on the hardware resources of the numerical system,
- Module II, specific for the application, independent on the hardware resources. To pass the product on other numerical systems it is enough to rewrite the first module.

The functional structure of the slave CANopen software is presented in Figure 14.

Module I is specific for the numerical system (phyCORE-mpc555, eZdsp-F2808, dsPIC-30F4011) and module II is common all three systems.

To implement the CANopen protocol it was used both the graphical programming and the classic (textual) programming.

6.3. Module I Implementation On The Ezdsp-F2808 Or Dspic-30f4011 Numerical Systems

The Simulink model visible structure of the slave CANopen communication node is presented in Figure 15.

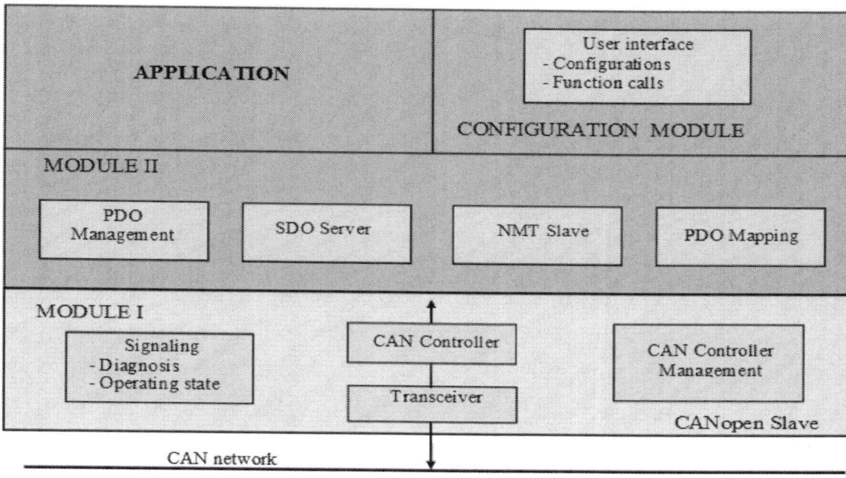

Figure 14. The functional structure of a CANopen slave

The CANOpen Message Receive (dsPIC30F4011 or eZdsp 2808) sub-system realizes the messages reception into the CANopen stack buffer. The messages are transmitted by the CANOpen Message Send (dsPIC30F4011 or eZdsp 2808) sub-system.

They are part of the Module I from the Figure 16. In the same module there is also the CANOpen Err & Run LED's sub-system which commands the two LEDs of the numerical system. The stack initialization and its periodical interrogation are realized by the Init CANOpen, and SW_TimerISR sub-systems.

The data transfer between the graphical and textual modules is made with global variables which are defined by the state flow chart. It is to mention that was necessary to interfere with the C-code generating files (Target Language Compiler – TLC) to obtain the necessary functionability.

An important aspect of the CANopen implementation is the generation of relative references of time to administrate the data transfer messages (timestamp) and the administrative data (node guarding, heartbeat).For this it was used a software which call both the CANopen stack and the timer with 1 ms period.

Module I implementation on phyCORE-mpc555 numerical system

The Simulink model for the CANopen node of the second numerical system is similar with the model from Figure 17 but eZdsp 2808 is changed with phyCORE-mpc 555. Thus, for a user which knows a model it is easy to operate with the other. The communication speed is established with the MPC555 Resource Configuration module.

Module II implementation of slave CANopen communication node

The graphical programming is operative and suggestive. It also has limits especially for the complex algorithms processing. In this case the programmer makes a compromise: hardware resources are realized with the graphical libraries and the complex algorithms processing are implemented with textual code lines. The Matlab/Simulink embraces such a combined programming.

Thus, the second module was implemented by a textual programming. The function call is realized with a 1KHz frequency by the SW_TimerISR sub-system. SDO services are assured by the object dictionary SDOResponseTable and by the functions Search_OD (WORD index, BYTE subindex), Send_SDO_Abort (DWORD ErrorCode) and Handle_SDO_Request (BYTE*pData). The functions Prepare_TPDOs (void) and TransmitPDO (BYTE PDONr) realize the administration of the data transmission messages between the numerical systems

The node initialization is realized by the function CANOpen_Init (BYTE Node_ID, WORD Heartbeat) and the communication network administration (NMT slave) are incorporate into the function CANOpen_ProcessStack(void).

The connections (mapping) between the data on the CAN communication bus can be static realized by the initialization function CANOpen_InitRPDO (BYTE PDO_NR, WORD CAN_ID, BYTE len,

BYTE *dat), CANOpen_InitTPDO (BYTE PDO_NR, WORD CAN_ID, WORD event_time, WORD inhibit_time, BYTE len, BYTE *pDat).

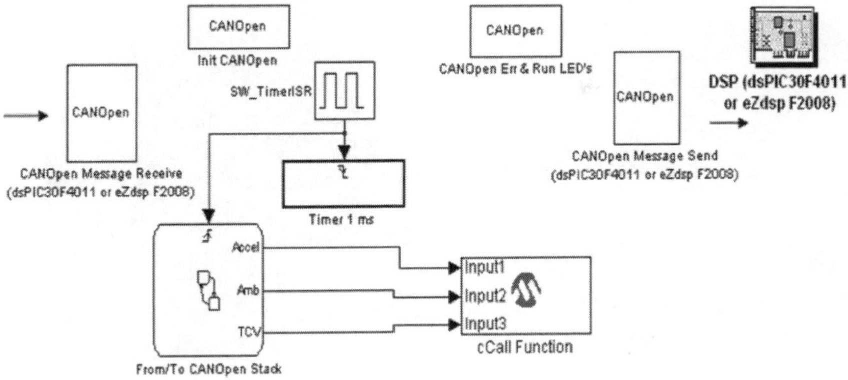

Figure 15. The Simulink model assigned to the slave CANopen communication node

6.4. Experimental Results

In Figure 16 is presented the hybrid electric vehicle model realized into the Energy Conversion and Motion Control laboratory of the Electrical Engineering Faculty from Iasi.

Finally several diagrams are presented highlighting the behaviour of the electric traction motor and the mechanical load emulator. It was considered a standard operating cycle UDDS (Urban Dynamometer Driving Schedule).

A velocity diagram UDDS cycle operation is shown in Figure 17-a. It is the speed reference for electric traction motor and the measured speed is presented in Figure 17-b.

The active current from electrical traction motor is shown in Figure 18-a. Also the mechanical load emulator is an electrical motor with torque control and the torque reference shown in Figure 18-b. InFigure 19 is presented the estimated torque from mechanical load emulator.

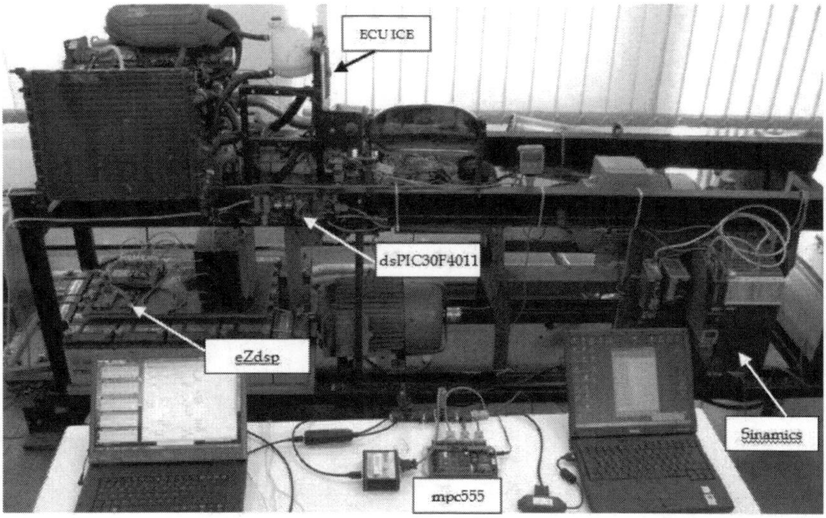

Figure 16. Hybrid electric vehicle experimental model

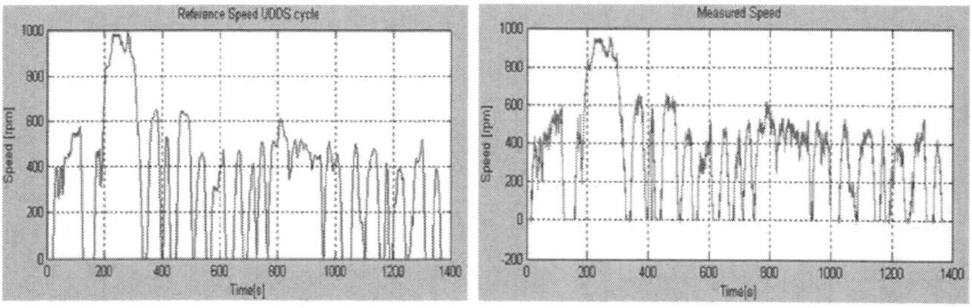

Figure 17. a) Reference speed for UDSS cycle b) Measured speed for electrical motor

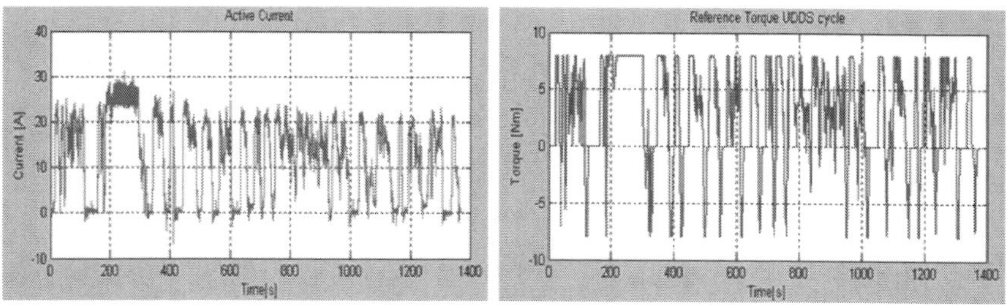

Figure 18. a) The active current from electric traction motor b) The reference torque for Sinamics system

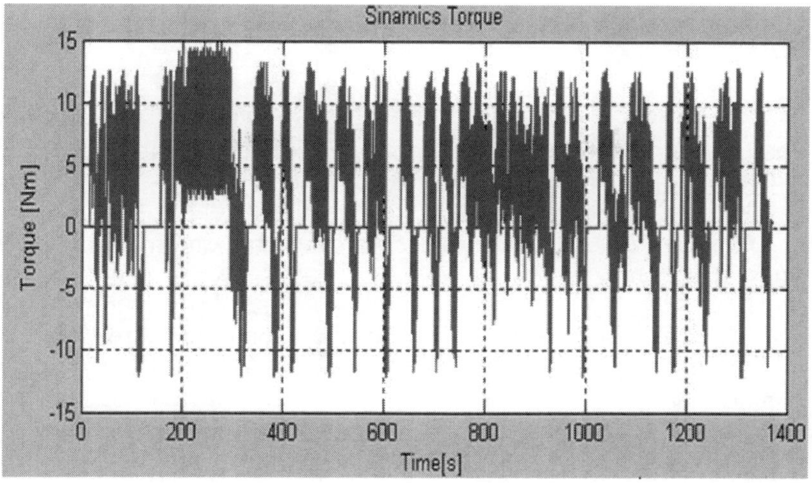

Figure 19. The estimated torque from mechanical load emulator

7. CONCLUSIONS

The hybrid electric vehicles are very complex dynamic systems and have an important number of interconnected electrical systems to achieve the required operating performances. Because of the complexity of the real time control for a hybrid electric vehicle it is necessary to integrate all the elements in a high speed CAN communication network to assure the distributed control of all the resources. For the hybrid electric vehicle experimental model is used a CANopen network with one master node and four slave nodes. The distributed system control with the CANopen protocol on a CAN bus permits the control of the electrical drives systems in safe conditions and with improved dynamic performances.

REFERENCES

1. K. C. Bayindir, M. A. Gozukucuk, A. Teke, 2011Acomprehensive overwiew of hybrid electric vehicle: Powertrain configurations, powertrain control techniques and electronic control units, Energy Conversion and Management, Elsevier, nr. 5213051313

2. CANopen, User Manual, Software Manual,2004PHYTE Technology Holding Company
3. V. R. Chacko, V. Z. Lahaparampil, V. Chandrasekar, 2005CAN based distributed real time controller implementation for hybrid electric vehicle, IEEE, 247251078039280905
4. C. C. Chan, 2002The state of the art of electric and hybrid vehicles, Proc. IEEE, 902247275
5. S. Comigan, 2002Introduction to the Controller Area Network (CAN), Texas Instruments Application Report, SLOA101-August 2002, 116
6. J. Duan, J. Xiao, M. Zhang, 2007Framework of CANopen protocol for a hybrid electric vehicle, Proceedings of the IEEE Intelligent Vehicles Symposium, Instanbul, Turkey, June 1315
7. M. Ehsani, Y. Gao, E. S. Gay, A. Emadi, 2005Modern Electric, Hybrid Electric, and Fuel Cell Vehicles CRC PRESS, Boca Raton London, New York, 0-84933-154-4
8. Emadi Ali,2005Hanbook of Automotive power electronics and MotorDrives, CRC PRESS, Taylor&Francis Group, LLC, 2005, 0-82472-361-9
9. A. E. Fuhs, 2009Hybrid Vehicles, CRC PRESS 2009, Taylor Francis Group, LLC,978-1-42007-534-2
10. L. Guzzella, A. Sciarretta, 2007Vehicle Propulsion Systems, Second Edition, Springer-Verlag Berlin Heidelberg, 978-3-54074-691-1
11. I. Husain, 2003Electric and Hybrid Vehicles Design Fundamentals, CRC PRESS, 0-84931-466-6
12. Gh. Livint, R. Gaiginschi, V. Horga, R. Drosescu, G. Chiriac, M. Albu, M. Ratoi, I. Damian, M. Petrescu, 2006Vehicule electrice hibride, Casa de Editura Venus, Iasi, Romania
13. Gh. Livinţ, M. Răţoi, V. Horga, M. Albu, 2007Estimation of Battery Parameters Based on Continuous-Time Model, Proceedings International Symposium on Signals, Circuits&Systems-ISSCS, July 12-13, 2007, Iasi, Romania, 613617IE.EE. Catalog 07EX1678C1-42440-969-1

14. Gh. Livinţ, V. Horga, M. Albu, M. Răţoi, 2007Evaluation of Control Algorithms for Hybrid Electric Vehicles, WSEAS TRANSACTIONS on SYSTEMS, 16January 2007, 1331401109-2777http://www.wseas.org

15. Gh. Livint, V. Horga, M. Ratoi, 2008Distributed control system for a hybrid electric vehicle implemented with CANopen protocol,-Part I, Bulletin of the Polytechnic Institute of Iasi, Tom LIV (LVIII), FASC. 4, 1223-813910191026

16. Gh. Livint, V. Horga, M. Ratoi, M. Albu, M. Petrescu, G. Chiriac, 2008Distributed control system for a hybrid electric vehicle implemented with CANopen protocol- Part II, Bulletin of the Polytechnic Institute of Iasi, Tom LIV (LVIII), FASC. 4, 1223-813910271032

17. Gh. Livint, V. Horga, M. Ratoi, I. Damian, M. Albu, G. Chiriac, 2008Advanced real rime control algorithms for hybrid electric vehicles optimization, CEEX Program, Simposion "Contributii Stiintifice ", UCP AMTRANS, Bucuresti, Romania, noiembrie 2008, 209214

18. Gh. Livint, V. Horga, M. Ratoi, M. Albu, G. Chiriac, 2009Implementing the CANopen protocol for distributed control for a hybrid electric vehicle, Proceedings The 8th International Symposium on Advanced Electromechanical Motion Systems, Lille, July 13CD., 978-2-91591-325-5EAN: 978-2-91 5913-26-5 IEEExplore, http://ieeexplore.ieee.org/xpl/ RecentCon.jsp?punumber=5234983

19. Gh. Livinţ, V. Horga, D. Sticea, M. Raţoi, M. Albu, 2009Electrical drives control of a hybrid electric vehicle experimental model, Proceedings of the 7th International Conference of Electromechanical and Power Systems, Editura PIM, 2009, vol. II, 2127ISBN vol II, 978-606-520-623-6, October 8-9, 2009, Iaşi, Romania,

20. Gh. Livinţ, V. Horga, D. Sticea, M. Raţoi, M. Albu, 2010Hybrid electric vehicle experimental model with CAN network real time control, in Advances in Eectrical and Computer Engineering, nr. 2., 2010, 1021081582-7445Stefan cel Mare University of Suceava, Romania

21. M. Petrescu, Gh. Livinţ, D. Lucache, 2008Vehicles dynamic control using fuzzy logic, Proocedings of 9th WSEAS International Conference on Automation and Information, pag. 488-493, 2008

22. Seref Soylu,2010Urban Transport and Hybrid Vehicles, Published by Sciyo, Janeza Trdine 9, 51000 Rijeka, Croatia, 978-9-53307-100-8

23. Siemens, Sinamics, S120 Control Unit and additional system components,2007Equipment Manual 03, Edition

24. D. Sticea, Gh. Livinţ, M. Albu, G. Chiriac, 2009Experimental stand for the dynamical cycle study of the batteries used on hybrid electrical vehicle, Proceedings of the 7th International Conference of Electromechanical and Power Systems, Editura PIM, 2009, vol. II, 172175ISBN vol II, 978-606-520-623-6, October 8-9, 2009, Iaşi, Romania,

25. E. Yamada, Z. Zhao, 2000Applications of electrical machine for vehicle driving system, Proceedings of the Power Electronics and Motion Control Conference (PIEMC), 313591364Aug. 15-18, 2000.

26. H. M. Westbrook, 2005The Electric Car, Developmrent and future of battery, hybrid and fuel-cell cars, The Institution of Electricl Engineers, London, 2005, 0-85296-013-1

27. F. A. Wyczalek, 2000Hybrid electric vehicles year 2000, Proceedings of the Energy Conversion Engineering Conference and Exhibit (IECEC) 35th Intersociety, 1349355July 24-28, 2000

Index